Springer Praxis Books

Astronomy and Planetary Sciences

Textbooks and monographs published in this series from 2013 onward are written for advanced undergraduate students in astronomy and the planetary sciences and advanced amateur astronomers. The editors insist on good readability and encourage new approaches to teaching astronomy and planetary sciences. Books published before 2013 serve a spectrum of readership: Some are at advanced amateur to advanced undergraduate level. Others are targeted at PhD students and researchers. Topics covered in the series include

- Astronomical telescopes and instrumentation
- Astronomical techniques, software and data
- Astrophysics, Astrochemistry, Astrobiology
- Solar system science (excluding the Earth sciences proper) and exoplanets
- Stellar physics and black hole astrophysics
- Galactic astronomy
- Extragalactic astronomy and cosmology

The books are well illustrated with line diagrams and photographs throughout, with targeted use of colour for scientific interpretation and understanding. Many feature worked examples or problems and solutions.

Sylvio Ferraz-Mello

Chaotic Dynamics in Planetary Systems

Springer

Sylvio Ferraz-Mello
Institute of Astronomy, Geophysics
and Atmospheric Sciences
University of Sao Paulo
São Paulo, Brazil

Springer Praxis Books
ISSN 2366-0082 ISSN 2366-0090 (electronic)
Astronomy and Planetary Sciences
ISBN 978-3-031-45815-6 ISBN 978-3-031-45816-3 (eBook)
https://doi.org/10.1007/978-3-031-45816-3

Originally published as "Caos e planetas: dinâmica caótica de sistemas planetários" by Editora Livraria da Física (2021)

Revised translation from the Portuguese language edition: "Caos e planetas: dinâmica caótica de sistemas planetários" by Sylvio Ferraz-Mello, © Editora Livraria da Física 2021. All Rights Reserved.

Cover illustration: This image, taken with the Atacama Large Millimeter/submillimeter Array (ALMA), in which ESO is a partner, shows the PDS 70 system, located nearly 400 light-years away and still in the process of being formed. The system features a star at its centre and at least two planets orbiting it, PDS 70b (not visible in the image) and PDS 70c, surrounded by a circumplanetary disc (the dot to the right of the star). The planets have carved a cavity in the circumstellar disc (the ring-like structure that dominates the image) as they gobbled up material from the disc itself, growing in size. It was during this process that PDS 70c acquired its own circumplanetary disc, which contributes to the growth of the planet and where moons can form. Credit: ALMA (ESO/NAOJ/NRAO) / Benisty et al. The PDS in this star's name stands for Pico dos Dias Survey, a survey that looked for pre-main-sequence stars at the Laboratório Nacional de Astrofísica, Brazil.

This Springer imprint is published by the registered company Springer Nature Switzerland AG
The registered company address is: Gewerbestrasse 11, 6330 Cham, Switzerland

Paper in this product is recyclable.

Adilce
my Love
my Life

Preface

The purpose of this book is the presentation of various dynamical systems that occur in the Solar System, in the systems of exoplanets, and in other celestial systems in which chaotic behavior is observed, with emphasis on the diagnostic techniques used to detect the occurrence of chaos. The discussion of each case presented in the book is based on theory, with special attention to the origin of the chaotic behavior and its phenomenological significance. The first chapter gathers some basic concepts that are presented in a simple way. The use of common language in the presentation implies a loss of mathematical rigor, but it is adopted so that the concepts can be understood by readers from the most diverse areas of study. For the same reason, some extensive descriptions that may seem superfluous to an advanced reader have been included. They can be useful for beginners and non-specialists. Readers interested in a presentation of these concepts with mathematical rigor should consult the many excellent books that exist on the subject and include more technical approaches to the topics of conservative chaotic dynamics. Many of them are cited in this book.

This book is written primarily for graduate and advanced undergraduate students. A first version of the book was published in Portuguese aimed at those students of my country that might feel more comfortable reading a book in their mother tongue than in a foreign language. This improved version, in English, is written to overcome this limitation and with the hope of giving, to a wider audience, an overview of an approach to Celestial Mechanics that has contributed in many ways to our understanding of the dynamics of planetary systems and which occupies a central place since the last decades of the last century.

When I was invited, some years ago, to write a friends' book (Dvorak and Lhotka, 2013) presentation, I wrote the following: "All books have a strong personal component in the topics that one chooses to include or not in it. This book is not an exception to this rule". This one is no exception to that rule either! The topics

The original version of the book has been revised. A correction to this book can be found at https://doi.org/10.1007/978-3-031-45816-3_4

included in the chapters on planetary systems dynamics are those that have been at the center of the research activities of the Solar System Dynamics group at the University of Sao Paulo over the past 40 years. It is not an Encyclopedia but rather a course book. The dynamics of planetary systems is a discipline of vast contours, and it is almost impossible to cover all topics of interest in a book centered on some current research topics. In fact, the fascination of this discipline lies in the immense amount of problems that can be studied in depth by extensively exploring the tools of Mathematics and the concepts of Physics.

The texts included in this book are based on courses on Chaos in the Solar System given at the graduate school of the Institute of Physics of the University of São Paulo in 1993 and 1997, on open courses on this subject given between 1995 and 2003 in La Plata, Zaragoza, Recife, and Rio de Janeiro, on courses on exoplanets taught, since then, at several international summer schools (El Leoncito, Cortina d'Ampezzo, Potsdam and Isle of Skye), in addition to many lectures.

Acknowledgements I thank Professor Iberê Luiz Caldas for proposing the course on Chaos in the Solar System, which is at the origin of this book. I thank Professors Rudolf Dvorak and Cristian Beaugé for their extremely helpful suggestions and comments. This book was written with the support of CNPq (Proc. 303540/2020-6) and FAPESP (Proc. 2016/13750-6 ref. Mission PLATO).

Reference

Dvorak, R., Lhotka, C.: Celestial Dynamics: Chaoticity and Dynamics of Celestial Systems. Wiley-VCH, Weinheim (2013)

São Paulo, Brazil Sylvio Ferraz-Mello

Contents

Chapter 1
Introduction: Conservative Chaotic Dynamics

1.1 Planetary Motion

The basic dynamical system of Celestial Mechanics is the motion of a planet around the Sun in the approximation called the two-body problem, that is, in an ideal planetary system in which there would only be the Sun and the planet. The solutions of this system are defined by Kepler's three laws:

- *Law of the elliptical orbits (1609)*—The planetary orbits are elliptical and the Sun is at one of the foci of the ellipses;
- *Law of equal areas (1609)*—The planets move in their orbits with such velocity that the radius vector connecting the planet to the Sun sweeps equal areas in equal times (see Fig. 1.1);
- *Harmonic Law (1619)*—The cube of the great axis of the ellipse traveled by the planet is proportional to the square of the time the planet takes to complete one revolution around the Sun.

The third law gives rise to one of the most important dynamical characteristics of planetary motion. It means that if we consider two planets with ellipses of different dimensions, their orbital periods will be different; for instance, if one of the planets is in an ellipse whose semi-axis is twice the semi-axis of the ellipse of the other, the time required for this planet to describe one complete revolution around the Sun will be $\sqrt{8}$ times larger than the other. Then, two planets moving in different orbits and which are at the initial moment in neighboring positions, do not remain indefinitely close to one another; the planet in the outermost orbit has a slightly longer orbital period and, therefore, lags behind the planet in the innermost orbit (see Fig. 1.2 left); if in one revolution the recorded delay is of an angle α, in two revolutions it is of 2α, in three revolutions of 3α, and so on.

For comparison, consider the basic system of linear physics, the harmonic oscillator, which governs the oscillations of a physical system in the immediate vicinity of an equilibrium solution. This system is characterized by an acceleration

S. Ferraz-Mello, *Chaotic Dynamics in Planetary Systems*,
Astronomy and Planetary Sciences, https://doi.org/10.1007/978-3-031-45816-3_1

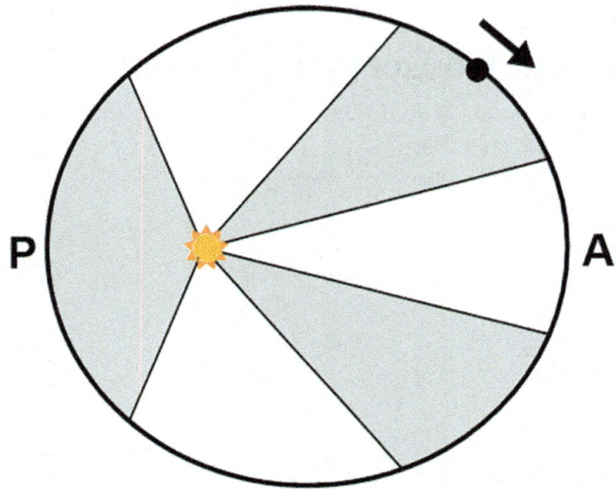

Fig. 1.1 Planetary motion. The velocity of the planet in the orbit is such that arcs bounding equal areas are traveled in equal times. The velocity is maximum in the perihelion (P) and minimum in the aphelion (A)

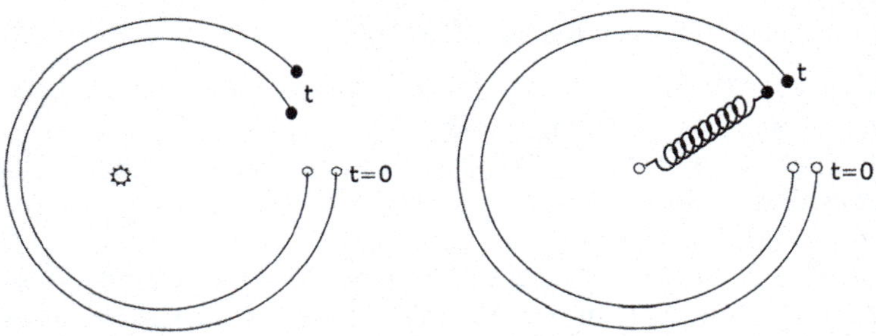

Fig. 1.2 *Left*: neighboring planetary motions. *Right*: neighboring motions of a mass attached to an ideal spring

proportional to the distance: $\ddot{x} = -kx$. When we solve this equation, we obtain oscillations whose frequency is directly related to the proportionality constant k. As Galileo discovered when studying the small oscillations of a pendulum, it does not matter whether the amplitude of the oscillation is larger or smaller, the period is always the same. Small oscillations are isochronous.

Let us consider the case of a planar oscillator, replacing x by a vector $\mathbf{r}$ in the plane[1]. We then have: $\ddot{\mathbf{r}} = -k\mathbf{r}$. The solutions are now concentric ellipses and, again, of the same period. Two distinct solutions with neighboring initial conditions remain close indefinitely. For instance, two solutions of different amplitudes which originate with equal phases (one in front of the other) remain indefinitely with

[1] In this book, bold characters are used to indicate vectors.

equal phases (see Fig. 1.2 right). In general, the first approximation of many Physics problems always has this characteristic of isochronism.

The type of behavior we observe in the harmonic oscillator is what is called, in Mathematics, stable motion (as defined by the mathematician Aleksandr Lyapunov). The isochronism of the solutions implies their stability. Therefore, due to Kepler's third law, the Keplerian motion is intrinsically different from the motion of a harmonic oscillator. The Keplerian motion is unstable (again, in Lyapunov's definition). If we consider two "Earths": the real Earth and a fictitious Earth which, at the initial instant, is 10 meters farther away from the Sun than the real Earth, the period of the fictitious Earth around the Sun is 0.003 seconds longer than that of the real Earth. That is, if the two start moving from the same phase (i.e., both in the same line passing through the Sun), after one year the fictitious Earth will be 0.003 seconds behind the real Earth, or, as the Earth travels in space with a speed of about 30 km/s, the fictitious Earth will be 90 meters behind the real Earth. This delay will grow in arithmetic progression: 0.006 seconds or 180 m at the end of the second year, 0.009 seconds or 270 m at the end of the third year, and so on. In one million years, the delay will be 50 minutes, and the fictitious Earth will be 90,000 km behind the real Earth. The proportional increase in the distance between the two Earths is one of the basic characteristics of classic instability.

1.2 Chaotic Motion of the Planets

The motion of the Earth around the Sun in the real Solar System, which, besides the Earth and the Sun, comprises seven more planets, is not only unstable but also chaotic (Laskar 1989; Sussman and Wisdom 1992).[2] This means that the motion of the planets of the Solar System presents an extreme sensitivity to the initial conditions and, besides the classical instability that grows linearly, an instability with approximately exponential growth also occurs. Figure 1.3 illustrates how this instability occurs. In order to observe this instability, it is necessary to consider all the planets moving from neighboring initial conditions, not only the Earth. The quantity $d(t)$, whose variation is shown in the figure, is a quantity composed of the variations of the geometric elements of the orbits of the 8 planets: eccentricities, inclinations, perihelions, and nodes, starting with a tiny initial variation (Laskar, 2017). We see that the distance $d(t)$ does not increase monotonically; on the contrary, there are many occasions when it decreases. But, in general, it grows more or less tenfold every 10 million years, thus exhibiting the geometric, or exponential,

[2] We say that a motion is chaotic when it presents extreme sensitivity to the initial conditions. The word "extreme" does not have a clear meaning. Maybe that is why several authors insist on it. Sometimes, it is replaced by "exponential". But, on the contrary, in this case, the word has a very precise meaning that, although it translates to a certain extent, what happens in a chaotic motion, is not exact and does not reflect the full complexity of these motions.

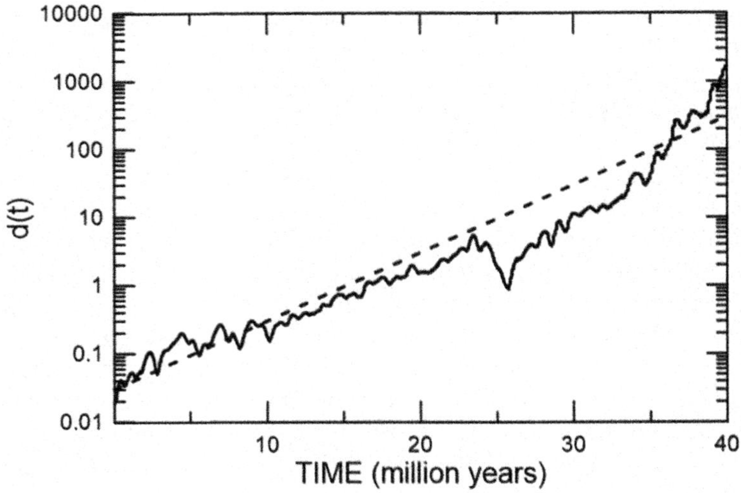

Fig. 1.3 Evolution of the distance between two solutions for the orbits of the 8 Solar System planets with neighboring initial conditions, measured by a quantity composed of their geometric orbital elements. The dotted line represents an exponential growth that multiplies the distance by 10 every 10 million years. Data source: Laskar (2017)

growth that characterizes the chaotic motion. This is what is called "extreme" sensitivity to initial conditions.

Thus, two solutions which, in the beginning, are at a small distance $d \sim 0.03$ (d represents the distance in phase space in arbitrary units) move apart and, after 40 million years, will be at a distance 10,000 times greater. The initial distance of 0.03 grows to about 300. A similar growth in distance between two neighboring orbits of two Earths, which started within 10 meters of each other, would take them, in 100 million years, to 100 million km (while growth due to ordinary instability would be only 9 million km).

This example shows what it means to say that there is an extreme sensitivity to initial conditions. It also allows us to distinguish the chaotic behavior from the simple ordinary instability in which the sensitivity of the solutions to the initial conditions exists but is not extreme (it is linear). In any case, the previous discussion contains some exaggerations. The growth of the distance between two neighboring solutions does not follow a perfect exponential. Do not think that the distance between the real Earth and a fictitious Earth could continue to increase tenfold every 10 million years forever. If this were the case, the distance would reach 10 billion km in 120 million years! That is impossible. In fact, the exponential behavior due to chaos only occurs while the two orbits are close enough; therefore, we cannot keep increasing the deviation indefinitely. After moving apart, the two orbits will follow independent and uncorrelated paths.

The chaos in our planetary system is relatively weak, and the evolution shown in Fig. 1.3 closely follows the patterns of idealized models. However, the motion of the

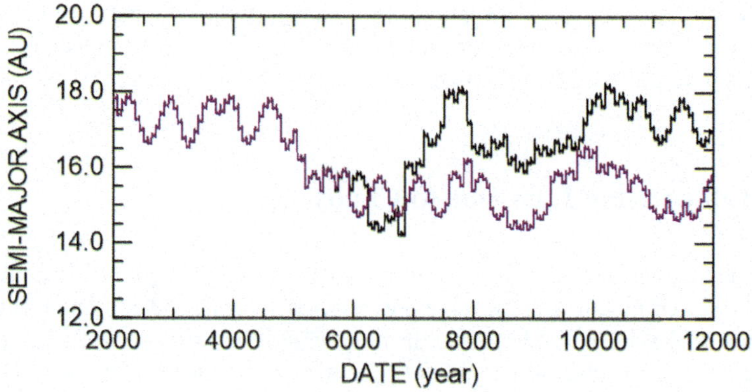

Fig. 1.4 Evolution of the semi-major axis in two solutions for the motion of the comet 1P/Halley which initially differ by one tenth of a thousandth of a degree in their longitudes

smaller bodies in the Solar System presents realities that do not accommodate these models well. One example is the variation of the semi-major axis of comet 1P/Halley in two solutions which initially differ by only one tenth of a thousandth of one degree in their longitudes. Initially, the two orbits are graphically indistinguishable. If we were interested in studying this period, we would have to use a graph showing specifically the difference between the two solutions on a logarithmic scale (we would have something like Fig. 1.3). In Fig. 1.4, only after some thousand years do we observe points where the two solutions do not coincide, but soon afterward, just before the year 6000, they become separate, and there is no longer any correlation between them. The cause of this separation must be a passage closer to one of the great planets, capable of exacerbating the growing separation between the two solutions.

The models currently used to determine the future position of the planets have limited validity due to the chaotic behavior of the solutions. For instance, the validity of the predictions of the latest version of the INPOP model of the Institute of Celestial Mechanics and Calculation of Ephemerides (IMCCE), a branch of the Paris Observatory, is 52 million years (Laskar, 2017). It is not expected that new models can go further. To extend the predictions to longer times, a better knowledge of the motion of other bodies, for example, the two largest asteroids, Ceres and Vesta, would be required. However, the motions of these asteroids present a chaoticity ten times greater than that of the inner planets, and it is impossible to have them well known. The impossibility of knowing their future orbits prevents us from knowing how they will affect the motions of the other planets and limits the models used to predict the motions of the planets. Certainly, from a practical point of view, not knowing where the planets will be 50 million years from now is not a problem, but unpredictability also applies to the past and knowledge of planetary orbits in the past is valuable in studies of geologic profiles and climate variations of past ages. The dating of samples is an important element in geology and paleoclimatology, and correlations with planetary movements have been one of the tools used to increase their accuracy. The chaos in the Solar System limits the use of them to the last 50 million years.

1.3 Information Loss Due to Chaos

Let us say that we know the position of the Earth and the other planets to an accuracy of 100 meters. Then, if we repeat the previous discussions and call one Earth A and the other B, we are not able to say, in fact, which one is closer to the real Earth. Since the motion is chaotic with the exponential divergence rate mentioned earlier, the Earths A and B will be 1000 km apart after 40 million years, 1 million km apart after 70 million years, etc. If we were to proceed by increasing the distance tenfold every 10 million years, we would find that the two Earths would be 100 million km apart in 90 million years. We have already said above that this is not so simple. In any case, the growth of the distance between Earths A and B is such that it is impossible to know, after a certain time, where the real Earth will be.

This example shows that a basic characteristic of chaotic systems is the loss of information. The exponential growth of the initial separation of 100 m leads to a separation so large that, after a certain time, we "lose" the Earth. We no longer know in which part of its orbit it is.

To measure the information loss, let us consider a chaotic system and suppose that two neighboring solutions of this system have an exponential divergence characterized by a tenfold increase every time T_{10}. Let us consider the problem of predicting its evolution from the current knowledge of the system using a numerical computer simulation. The current state of the system is given to the computer by numbers. Let us say, for instance, that the number given to the computer is composed of six digits. This fact implies that the computer gets the initial position of the system with an indeterminacy of the order of the unit represented by the last digit. As this system is chaotic, this indeterminacy will grow in the same proportion in which diverge two different orbits whose initial conditions are given by the same six digits, the seventh digit being different. After a time T_{10}, the indeterminacy in the position of the system will be about 10 times larger than the initial indeterminacy, that is, the sixth digit of the result will no longer have any meaning; after $2T_{10}$, the indeterminacy will be about 100 times the initial indeterminacy, that is, the fifth digit of the result will no longer have any meaning; and so forth. This means that the transfer of our knowledge about the initial state of a chaotic phenomenon to a later date is done with a loss of information of one digit every T_{10}. If we want to measure the information loss using the usual unit for measuring information, the bit, it suffices that, instead of considering the decay time T_{10}, we consider the doubling time T_2 ($T_2 = [\log_{10} 2]\, T_{10} \sim 0.3\, T_{10}$). At each T_2, we lose 1 bit of information. In the case of our planetary system, one bit is lost every 3 million years.

This loss of information should not be confused with the ordinary propagation of errors introduced by the computer and the numerical integration program used to perform the simulation. Both can be assumed perfect. The processes of ordinary error propagation are quite well known and are given by laws in which the error always propagates proportionally to a power of t (less than 3/2) depending on the type of error considered.[3] To remove any doubt in this respect, we can consider two similar problems, one chaotic and the other non-chaotic, and study the solutions that emerge from the same initial conditions. In one of the cases, the error grows with a small power of t, while in the other, it grows exponentially. In the non-chaotic case, there is also a loss of information due to the propagation of errors, but very slow.

We should also not confuse the unpredictability of chaotic phenomena with the unpredictability of random phenomena. In random phenomena, unpredictability is due to the existence of parameters whose evolution is not causal, as in a dice game where the result of each throw does not depend on the result of the previous throw. In chaotic phenomena, causality is present. The value of the solution at each instant of time is determined by the value of the solution at the previous instant of time. The loss of accuracy occurs because of the complexity of the dynamics of the system. If the phenomenon were known at one moment with infinite precision, we could predict its future evolution.

1.4 Chaos in the Rotation of Hyperion

While in planetary motion, the consequences of chaos manifest themselves over millions of years and only affect the use of models in support of geological and paleoclimatic studies, there are examples where the effects are much more rapid. An example of more immediate consequences is the rotation of Hyperion. Hyperion is a satellite of Saturn (discovered in 1848 by W.C.Bond, G. P.Bond, and W.Lassel) that orbits around the planet, with a period of 21 days. Its shape is quite irregular, with about 190 km along the direction in which it is more elongated, and 145 km and 115 km along axes perpendicular to this direction (see Fig. 1.5). One of the problems with this satellite is that its axis of rotation is very close to the smallest of the three axes mentioned above. The main perturbation affecting the rotation of Hyperion is the gravitational attraction of Saturn. Saturn's attraction is stronger on the hemisphere facing Saturn than on the other; this difference in the attraction on one hemisphere and on the other creates a torque that causes a precession of the rotation axis of the satellite. To study the rotational motion of Hyperion, we must consider the Euler equations governing the evolution of its spatial orientation or

[3] The errors introduced in the simulation are due to the truncation of the results of the arithmetic operations to the number of digits that the computer uses to represent a real number and to the truncation of the series used to represent the solution, which replaces a continuous trajectory of continuous derivatives by a sequence of juxtaposed small arcs, with discontinuities in the higher order derivatives.

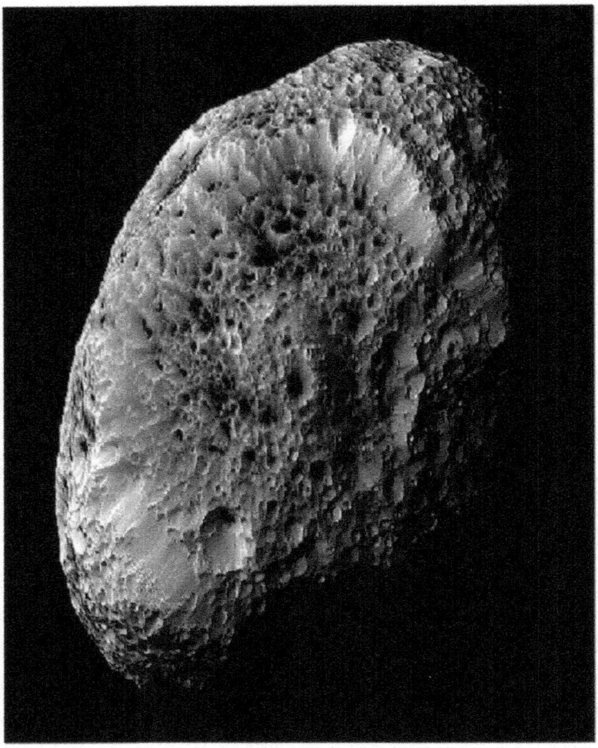

Fig. 1.5 Hyperion. Image taken by the Cassini spacecraft (Credit: NASA/JPL-Caltech/ Space Science Institute)

attitude[4]. Let us suppose that, at a given instant, the values of the angles fixing the orientation of Hyperion are known with a certain precision. As the rotation of Hyperion is chaotic, the error in the determination of its orientation propagates with an almost exponential law. In this case, the growth rate of the errors is relatively large; they are multiplied by 10 every 92 days (approximately; see Sect. 2.13). The satellite Hyperion was visited by two space probes, Voyager I and Voyager II, which passed close to Saturn and photographed this object. The two probes were launched from Earth on August 20 and September 5, 1977, respectively, but followed slightly different paths so that they passed by Saturn, one on November 12, 1980, and the other on August 25, 1981, 286 days later. That is enough time for the errors in the position determined in the first visit to increase about 1000 times. Thus, even if the rotational velocity of the satellite was perfectly known and the angles fixing the spatial orientation of the satellite determined with a precision of 0.1°, a numerical

[4] In Astrodynamics, the word "attitude" is used to designate the spatial orientation of a solid. It can be known, for instance, by fixing the three Euler angles.

simulation using these data as initial conditions would not allow us to know what the orientation of the satellite would be at the time of the arrival of the Voyager II probe. The error of 0.1° in the initial orientation would grow to almost 100°, which, in practice, is equivalent to not knowing which would be the orientation of the satellite upon the arrival of the Voyager II space probe.

This example allows us to better characterize the kind of phenomena we want to study. We want to study phenomena whose behavior is such that distinct evolutionary paths beginning with almost the same conditions diverge enormously until they become completely different.

1.5 Gain of Information Thanks to Chaos

Let us consider another aspect of this problem. To make it simpler to explain, let us replace Hyperion with the Moon and consider any parameter fixing the attitude of the Moon relative to the Earth. This parameter will have a periodic evolution. If we follow this parameter for a long time, we will observe some modulations in its periodic variation. In general, in order to obtain information about the physical conditions prevailing in a given astronomical phenomenon, it is necessary to observe it under various conditions. But in Astronomy, contrary to what happens in Physics, we cannot change the conditions under which the phenomena occur: we can only observe them in the conditions under which they do occur. We have to wait for their spontaneous variations. When the information we receive is characterized as an almost periodic signal, in which one piece is almost the same as the next, many cycles are necessary to accumulate an important amount of information. If we study the motion of the Moon for 28 days, and then wait a year and study it for another 28 days, the gain will be minimal because things are reproducing themselves in an almost identical way; we will be able to see if the period of rotation remained constant or how much its (tiny) variation was, but we will have no more information than this. But in the case of Hyperion, things are different. At each moment that we observe it, we are gathering new information. And this is what happened with Hyperion. As the physics of the rotation of a body is well known and as the forces acting on Hyperion (mainly due to Saturn) are well known, it was possible to build evolutionary models in which many parameters are involved, and, by comparing the evolution of the models with the observed evolution of the satellite, to determine in a precise way several of the parameters involved in the model. Thus, although the time during which this satellite was studied was short, thanks to its irregularity, it was possible to make a good determination of its physical parameters, namely its moments of inertia.

Here we can reproduce a comment made by some stellar astrophysicists dealing with variable stars. For a century, the regular variable stars, whose brightness varies periodically, have been exhaustively studied, while their siblings with irregular brightness variation have been less studied. A greater wealth of information about the stellar interior should therefore be expected from the irregular variable stars because they can be observed at different phases of their evolution.

1.6 Conservative Mechanics

The equations governing celestial motions are essentially of the type

$$m\ddot{\mathbf{r}} = \mathbf{f}_{\text{applied}} \tag{1.1}$$

(the change in the linear momentum of a body is equal to the applied forces) and, in the case of a solid body,

$$\dot{\mathbf{s}} = \mathcal{M} \tag{1.2}$$

(the change in the angular momentum is equal to the torque of the applied forces).

These equations express the two physical laws which give the evolution of the position and attitude of a celestial body. The forces applied are the gravitational forces. For example, in the case of the interaction of two bodies: $\frac{GMM'}{r^2}$ where G is the constant of universal gravitation, M and M' the two attracting masses, and r the distance that separates them.

The most important forces acting in the motion of celestial bodies, in general, depend only on the relative positions of the bodies. They do not depend upon their velocities. Moreover, in general, they are derived from a potential, and it is possible to write them as a gradient of a given function.[5] Thus, in isolated systems, we can always find for these equations a conservation law (or first integral), the energy integral.

Problems in which dissipative forces act are also important in Solar System Mechanics. Examples are systems in which friction occurs, as in bodies moving in a gaseous medium that offers resistance to motion (such as satellites in the upper layers of the Earth's atmosphere) and in the case of extended viscous bodies deformed by tidal forces (Ferraz-Mello et al., 2020). Chaos effects are observed in both conservative and dissipative systems. Conservative systems are the only ones considered in this book.

1.7 Two-Body Problem

The equation governing the motion of a celestial body (abstracting its volume and considering only the motion of its center of gravity) is a second-order differential equation. The variable $\mathbf{r}$ is a vector. Therefore, the order of the system is 6. If we introduce the velocity $\mathbf{v}$ as an additional variable, independent of the previous one,

[5] In this book, we follow the Physics laws and write, e.g., $\mathbf{F} = -\text{grad}\ W$. The omission of the minus sign, as usually done in Mathematics, is formally correct, but W can no longer be identified to a potential energy.

and decompose the system into first-order equations, the result is a system of two first-order vector equations:

$$\dot{\mathbf{r}} = \frac{d\mathbf{r}}{dt} = \mathbf{v}, \qquad \dot{\mathbf{v}} = \frac{d\mathbf{v}}{dt} = \frac{\text{applied force}}{\text{mass}}. \tag{1.3}$$

If we are studying the motion of a single body, we have 6 scalar equations, but in the simplest real case, we have at least two mutually attracting bodies. Then, instead of the above equations, we have to consider the system

$$\frac{d\mathbf{r}_i}{dt} = \mathbf{v}_i, \qquad \frac{d\mathbf{v}_i}{dt} = \frac{\mathbf{f}_i}{m_i} \tag{1.4}$$

with $i = 1, 2$ and $\mathbf{f}_1 = -\mathbf{f}_2$. We then have 12 scalar differential equations. In the two-body problem, it is possible completely integrate the system of equations. Instead of considering the two position vectors $\mathbf{r}_1$ and $\mathbf{r}_2$, we can introduce a reference system fixed on one of them and refer to it the position of the other. One only caution is necessary (but it is a fundamental one): the law of Physics used here is Newton's law $\mathbf{f} = m\mathbf{a}$, and this law is only valid if the used reference system is inertial. It is not correct to use Newton's law in a moving reference system connected to one of the bodies. We then use a reference system with the center at the barycenter of the two masses, which is inertial since the motion of the barycenter of an isolated system is inertial. The equations obtained from the application of Newton's law in this inertial frame of reference can then be transformed into the moving reference system through simple geometric transformations. By doing this, we obtain a set of equations for the vector $\mathbf{r}$ that has its origin in one body and its end in the other. In fact, this transformation is very simple: we just introduce $\mathbf{r} = \mathbf{r}_2 - \mathbf{r}_1$ and $\mathbf{v} = \mathbf{v}_2 - \mathbf{v}_1$. It is then enough to subtract these equations from one another to obtain the equations of relative motion. Because of the aforementioned characteristic, that the gravitational interaction depends only on the relative positions of the bodies (it does not depend on $\mathbf{r}_1$ or $\mathbf{r}_2$ alone), the reduction of the initial problem to the equations of relative motion is trivial and results in 2 vector equations:

$$\frac{d\mathbf{r}}{dt} = \mathbf{v}, \qquad \frac{d\mathbf{v}}{dt} = \frac{\mathbf{f}_2}{m_2} - \frac{\mathbf{f}_1}{m_1}. \tag{1.5}$$

It is easy to verify that the motion governed by these equations conserves the total angular momentum, that is,

$$\mathbf{r} \wedge \mathbf{v} = \mathbf{c} = \text{constant}. \tag{1.6}$$

This means that the velocity and relative position must always be perpendicular to the vector $\mathbf{c}$. Given the initial conditions $\mathbf{r}_0$ and $\mathbf{v}_0$, the specific relative angular momentum $\mathbf{c}$ can be determined from

$$\mathbf{c} = \mathbf{r}_0 \wedge \mathbf{v}_0. \tag{1.7}$$

As the vector **c** is constant, we conclude that the motion is planar and occurs in a plane perpendicular to **c**. If the motion is planar, we can further simplify its equations by choosing the plane of motion as the principal plane of the reference system and working with only 4 scalar equations (the other two, fixing the z-coordinate and the velocity along the third axis, are no longer necessary).

Since the system is isolated, the 4 remaining variables must satisfy two conservation laws:

- the modulus of the angular momentum $c = |\mathbf{c}|$ is constant;
- the energy is conserved. E is constant.

That is, there are two functions $c(\mathbf{r}, \mathbf{v})$ and $E(\mathbf{r}, \mathbf{v})$, whose values, along each solution, remain constant and equal to their initial values. Thanks to these two equations, two more variables can be eliminated, and the resulting system has only two equations. Let us say that, using polar coordinates, they are written as

$$\dot{r} = f(r, \phi); \qquad \dot{\phi} = g(r, \phi). \tag{1.8}$$

As the time does not explicitly appear on the right-hand side of these equations, and $g(r, \phi)$ never vanishes (the variation of the angle ϕ is monotonic), we can use ϕ as the new independent variable and eliminate the time. The resulting equation is:

$$\frac{\mathrm{d}r}{\mathrm{d}\phi} = \frac{f(r, \phi)}{g(r, \phi)}, \tag{1.9}$$

whose solution is the equation of the followed trajectory. Thus, the problem of the two bodies is completely solved. We can remember that if the forces applied result from the Newtonian gravitational attraction of the two bodies, the trajectory $r(\phi)$ will be an ellipse, a parabola, or a hyperbola with the Sun occupying one of the foci of the conic. Figure 1.6 shows the hyperbolic orbit of the interstellar object 'Oumuamua and the inner planets, as seen from a point above the plane of their orbits (the Sun appears offset because, in the oblique projection of an ellipse, the focus is not projected onto the focus of the projection).

In summary: We begin with 12 equations. Then, implicitly using the physical law saying that "in every isolated system (that is, one in which no external force acts) the motion of the barycenter, with respect to an inertial reference system, is rectilinear and uniform", we place the origin of the reference system at the barycenter. With this, we have six algebraic relations which define that the barycenter is at the origin of the system and that it does not move:

$$m_1\mathbf{r}_1 + m_2\mathbf{r}_2 = 0; \qquad m_1\mathbf{v}_1 + m_2\mathbf{v}_2 = 0. \tag{1.10}$$

Then, we use the law of conservation of angular momentum. Two of the relations given by this law—those fixing the direction of the angular momentum—are used to show that the motion is planar and that, therefore, the plane of motion can be used as the fundamental plane of the reference system, reducing the system from

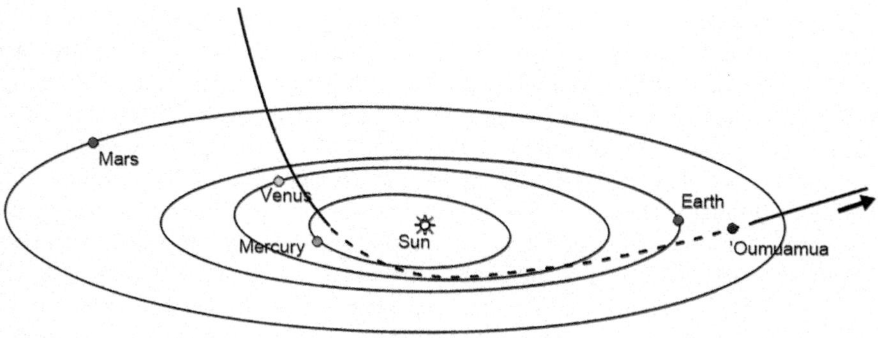

Fig. 1.6 Orbits and positions of the object 'Oumuamua and the inner planets in October 2017 cf. NASA/JPL-Caltech. 'Oumuamua approached the Sun in a hyperbolic orbit, indicating that it is an object not connected to the Solar System, perhaps coming from the vicinity of another star. The dotted line indicates the arc below the plane of the Earth's orbit (Southern hemisphere)

6 equations to 4. Then, we use the conservation of the modulus of the angular momentum together with the conservation of the energy to reduce the system to two equations. This leaves us with two equations with time-independent right-hand sides (that is, an autonomous system of two ordinary differential equations) that can be solved. After solving the equations, it is enough to repeat the sequence of operations backward to obtain $\mathbf{r}_i$ and $\mathbf{v}_i$ as functions of the initial positions and velocities, and of the time.

1.8 Flow Incompressibility (Liouville)

Let us return to the equations governing the interactions of several bodies:

$$\dot{\mathbf{r}}_i = \mathbf{v}_i; \qquad \dot{\mathbf{v}}_i = \mathbf{f}_i/m_i. \tag{1.11}$$

Note the following: the right-hand sides of the equations giving the $\dot{\mathbf{r}}_i$ do not depend on the positions $\mathbf{r}_i$, only on the velocities. Similarly, if we consider only forces $\mathbf{f}_i$ that do not depend on velocities, we have an analogous property in the equations giving $\dot{\mathbf{v}}_i$: they do not depend on the velocities $\mathbf{v}_i$. This simple-looking fact has a consequence that is fundamental in the behavior of chaotic solutions. Note that if we calculate the divergent of the vector $\begin{pmatrix} \dot{\mathbf{r}}_i \\ \dot{\mathbf{v}}_i \end{pmatrix}$ in the 2N-dimensional space formed by the $\mathbf{r}_i$ and $\mathbf{v}_i$, we have:

$$\operatorname{div}\begin{pmatrix} \dot{\mathbf{r}}_i \\ \dot{\mathbf{v}}_i \end{pmatrix} = \sum\{\frac{\partial}{\partial \mathbf{r}_i}\dot{\mathbf{r}}_i + \frac{\partial}{\partial \mathbf{v}_i}\dot{\mathbf{v}}_i\} = 0 \tag{1.12}$$

(the sum extending over all components of $\mathbf{r}_i$ and $\mathbf{v}_i$). That is, the divergent of the flow rate is null. This result means that, in this space, the solution bundles form a stream that flows as an incompressible fluid. A stream in this space can neither narrow nor widen. If it narrows in one direction, it must widen in another to preserve the measure of one set enclosed by it. This result is due to Joseph Liouville.

A first consequence of this property has already appeared in the solution of the two-body problem. That problem was stated as a system of 12 ordinary differential equations with time-independent right-hand sides completely integrated using only 10 conservation laws (or first integrals) when 11 should have been necessary. In fact, when we reached the last pair of equations, in r and ϕ, the possibility of integration and obtention of the equation of the trajectory was a direct consequence of the incompressibility of the flow.

Let us illustrate this with a simple problem of order 2. Consider two ordinary differential equations

$$\dot{x}_1 = \Phi_1(x_1, x_2); \qquad \dot{x}_2 = \Phi_2(x_1, x_2). \tag{1.13}$$

If the flow of $x_1(t), x_2(t)$ is incompressible, the divergent of the flow is zero:

$$\mathrm{div}\begin{pmatrix}\Phi_1\\ \Phi_2\end{pmatrix} = \frac{\partial \Phi_1}{\partial x_1} + \frac{\partial \Phi_2}{\partial x_2} = 0 \tag{1.14}$$

and then

$$\frac{\partial \Phi_1}{\partial x_1} = -\frac{\partial \Phi_2}{\partial x_2}. \tag{1.15}$$

But, this equation is the condition for the existence of a function $G(x1, x2)$ such that

$$\frac{\partial G}{\partial x_2} = \Phi_1; \qquad \frac{\partial G}{\partial x_1} = -\Phi_2 \tag{1.16}$$

and

$$\frac{\mathrm{d}G}{dt} = \Phi_1\dot{x}_2 - \Phi_2\dot{x}_1 = 0. \tag{1.17}$$

In the usual language of Mathematical Analysis, we say that the differential form

$$\mathrm{d}G = \Phi_1(x_1, x_2)\mathrm{d}x_2 - \Phi_2(x_1, x_2)\mathrm{d}x_1 \tag{1.18}$$

is exact. Its integration allows us to obtain the function G.

The incompressibility—or measure preservation, as this fact is usually referred to—is the determining factor in the way that chaotic solutions evolve in conservative systems.

1.9 Laws of Conservation (First Integrals) and Complete Integration

We will resume the end of the two-body problem study, but we will present it in a more general way. The system of four scalar differential equations which govern the motion in its plane can be written: $\mathrm{d}\mathbf{x}/\mathrm{d}t = \mathbf{f}(\mathbf{x})$ where $\mathbf{x}$ is a four-dimensional vector. But the four components of $\mathbf{x}$ are not independent. They have to satisfy the two relations given by the laws of conservation of angular momentum and energy. We can use them to eliminate two of the components of $\mathbf{x}$ from the equations, for example, x_3 and x_4, substituting them with c and E. To do this, we take the expressions of the two conservation laws: $c = c(x_1, x_2, x_3, x_4)$ and $E = E(x_1, x_2, x_3, x_4)$ and solve these equations for x_3 and x_4. Then, we substitute the results $x_3 = x_3(x_1, x_2, c, E)$ and $x_4 = x_4(x_1, x_2, c, E)$ in the given differential equations, which become

$$\frac{\mathrm{d}x_i}{\mathrm{d}t} = g_i(x_1, x_2, c, E) \tag{1.19}$$

where the g_i $(i = 1, 2)$ are the functions that result from the f_i when we introduce the above expressions for x_3 and x_4. But we no longer need all 4 equations because, when x_1 and x_2 are known, the equations expressing the conservation laws allow us to calculate x_3 and x_4. Thus, only the equations with the subscripts $i = 1, 2$ interest us. The incompressibility of the flow guarantees the integrability of the resulting system.

In practice, as the functions g_i do not explicitly depend on time, we can eliminate time from the two resulting equations. For instance, if $g_2 \neq 0$ over the whole interval under consideration, the function $x_2(t)$ is monotonic (as occurs for the position angle ϕ in the problem of the two bodies) and, therefore, x_2 can be introduced as a new independent variable so that one can write

$$\frac{\mathrm{d}x_1}{\mathrm{d}t} = \frac{\mathrm{d}x_1}{\mathrm{d}x_2}\frac{\mathrm{d}x_2}{\mathrm{d}t} \tag{1.20}$$

and therefore,

$$\frac{\mathrm{d}x_1}{\mathrm{d}x_2} = \frac{\mathrm{d}x_1}{\mathrm{d}t} \Big/ \frac{\mathrm{d}x_2}{\mathrm{d}t} = \frac{g_1}{g_2} = g(x_1, x_2, c, E); \tag{1.21}$$

that is, the solution of the system is reduced to a single equation and with its integration, we conclude the complete integration of the system.

1.10 A Non-integrable System: The 3-Body Problem

If instead of considering two bodies, we had considered three bodies, we would have 18 equations: 9 for the coordinates of the three bodies and 9 for the components of their velocities. Taking these coordinates referred to the barycenter, we have the relations

$$m_1\mathbf{r}_1 + m_2\mathbf{r}_2 + m_3\mathbf{r}_3 = 0; \qquad m_1\mathbf{v}_1 + m_2\mathbf{v}_2 + m_3\mathbf{v}_3 = 0; \tag{1.22}$$

which allow us, as in the previous case, to take one of the three bodies as the origin of a new reference system and eliminate 6 equations. There is still the angular momentum integral, which no longer has the simple interpretation of planar motion that it had in the two-body problem, but still allows us to eliminate 3 of the equations, reducing them to 9. At last, the energy integral allows us to reduce one more, to 8. Another reduction is still possible: it is the so-called node elimination, discovered by Jacobi, which comes from the fact that it is possible to choose the variables used in such a way that one of them (which expresses the position at which the plane of the three bodies intersects the reference plane) can have its equation separated from the others. Thus, the order of the system of differential equations can be reduced to 7. Since also in this case, the vector flow defined by the right-hand sides of the given equations is incompressible, we would need six more first integrals or conservation laws to solve the reduced system. Do they exist or not? This is a very important question and is closely associated with the subject studied in this chapter.

Do new conservation laws not exist, or do we not know how to calculate them? Do we stop at a system of 7 equations because of our ignorance? Could it be that with a better command of calculus, we would discover other conservation laws besides those connected with the motion of the barycenter, the constancy of angular momentum, and energy? This question was formulated for the first time at the end of the nineteenth century. In fact, the question was asked in a slightly different way. There was no reason to believe that it was not mathematically possible to continue the process until reduction to one final equation. Therefore, it was a matter of searching for the missing laws.

This problem was faced by several astronomers and mathematicians. The main results were obtained by Heinrich Bruns and Henri Poincaré. Bruns showed that it is not possible to obtain a new conservation law in the form of an algebraic function of positions and velocities. Poincaré's result was more restrictive. He showed that there is no other law of conservation that is an analytic function. If eventually such a function existed, it would necessarily contain singularities. That is, in the 3-body problem, we have the 10 conservation laws already discussed, the elimination of the node, and let us be glad with them. It is not possible to find any other conservation law that is analytic and that solves this problem. We then conclude that in this problem, it is not just a matter of not knowing how to perform this reduction; it is that, in fact, there is no new globally valid conservation law.

1.11 Integrability

In the three-body problem, we exhaust the conservation laws of the problem without solving it. The consequence is that it is not possible to write the solutions of the problem in a global form. That is, it is not possible to obtain a function that expresses $\mathbf{r}$ as a function of the initial conditions $(\mathbf{r}_0, \mathbf{v}_0)$ and of time. It is only possible to have such a function when all the necessary reduction operations can be performed. If we cannot do it, we are not able to know the positions as functions of initial conditions and time in a global way (that is, for all times and all initial conditions).

Locally, however, the solutions exist! Given the system of differential equations, $\frac{d^2\mathbf{r}_j}{dt^2} = \frac{\mathbf{f}_j}{m_j}$, we can numerically, using a computer, calculate its solution. It is enough to give the initial conditions—the initial positions and velocities of the three bodies—and from these conditions, the system will evolve. Some existence theorems in the theory of Ordinary Differential Equations (Peano, Lipschitz) guarantee it. Peano's theorem constructs the solution in the same way as when we calculate them by means of numerical integration: the solution is calculated piecemeal, with very small steps. These theorems say more or less the following: if the right-hand sides of the above equations are continuous and differentiable functions, there is a solution passing through the given initial condition, and this solution is unique (the actual condition is, in fact, less strict than differentiability, but this level of rigor is beyond our purposes here). This theorem gives rise to the question: How is the system non-integrable? Doesn't Peano's theorem say that it is possible to start from an initial condition and build the solution? This question allows us to make clear what was said before. When we say that the system is not integrable, we are not saying that it cannot be integrated, by a certain time, from a given initial condition. When we say that one system is non-integrable, we are only saying that it is not possible to write a globally valid solution to it.

The reasons for this impossibility will become clear in the following sections. When we go from the problem of the motion of two material points interacting gravitationally to an analogous problem, but with three interacting masses, we go from a problem that is integrable to a problem that is not integrable. And this makes a big difference. In an integrable system, chaotic solutions do not occur. In such systems, what always happens is that if we take two neighboring solutions that start far from a singular point of the system, and consider the distance between them over time, this distance may grow, but it will almost always be a more or less linear growth. If the solution is searched using numerical simulations on a computer, the phenomenon of rapid loss of information (i.e., rapid loss of significant digits) that occurs in the cases of chaotic solutions does not occur in the solutions of integrable systems. There will be propagation of errors, but it will be an ordinary propagation, with errors growing in time but limited by a power law, and no more than that.

1.12 Surfaces of Section. Poincaré Maps

Let us consider the integrability problem of a conservative mechanical system with 2 degrees of freedom:

$$\begin{aligned}\frac{dx}{dt} &= u\\ \frac{dy}{dt} &= v\\ \frac{du}{dt} &= -\frac{\partial U}{\partial x}\\ \frac{dv}{dt} &= -\frac{\partial U}{\partial y}\end{aligned} \tag{1.23}$$

where $U = U(x, y)$ is a potential dependent only on the position $\mathbf{r} \equiv (x, y)$ and independent of time and velocity $\mathbf{v} \equiv (u, v)$. This system conserves the energy:

$$E = E(x, y, u, v) = \frac{1}{2}(u^2 + v^2) + U(x, y). \tag{1.24}$$

Let us consider the problem of knowing whether this system is integrable or not. We have four equations and one conservation law. For the system to be integrable, there must be at least one more conservation law. Is there, in this system, some other conservation law that we are not aware of, which would allow us to obtain a complete solution for the system?

The process that we now describe was first described by Poincaré. We can initially use the law of conservation of energy and eliminate one of the variables. Let us take the law in the form $E = E(x, y, u, v)$ and suppose that it is possible to solve this equation with respect, for example, to v. We can then solve the equation $E = E(x, y, u, v)$ algebraically and write its solution as a function $v = \Omega(x, y, u; E)$. We replace this function in the right-hand sides of the given differential equations and obtain a system reduced to three differential equations

$$\begin{aligned}\frac{dx}{dt} &= u\\ \frac{dy}{dt} &= \Omega(x, y, u; E)\\ \frac{du}{dt} &= -\frac{\partial U}{\partial x}\end{aligned} \tag{1.25}$$

The fourth equation is no longer necessary since, given the energy E and knowing the solutions $x(t)$, $y(t)$, $u(t)$, the last variable $v(t)$ is automatically determined by the equation $v = \Omega(x, y, u; E)$.

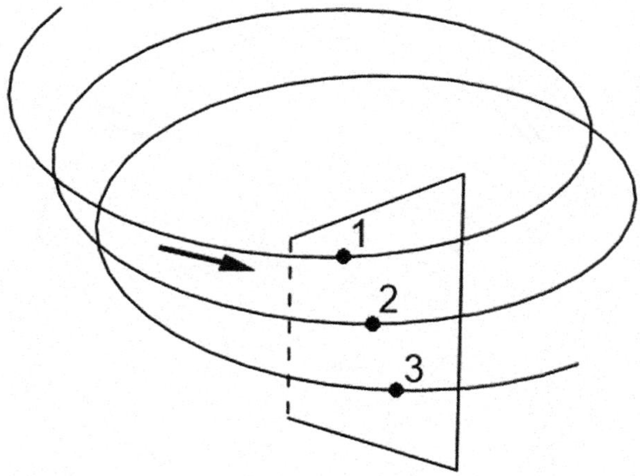

Fig. 1.7 Construction of the consequents in a Surface of Section

The problem has become three-dimensional, which allows us to try to graph the trajectory that starts from an initial point x_0, y_0, u_0, and that has a fixed energy E. To solve the difficulty inherent in the fact that we are trying to see a three-dimensional trajectory, we can cut the figure by planes (or other surfaces) that are effectively crossed by the solutions and try to represent the entire solution by marking on these planes the points where the trajectory cuts them. This technique is called the construction of a Surface of Section, and its result is a Poincaré map (or first return map). It is a map because to each point corresponds a unique solution, and each point perfectly determines the next one; Poincaré called "consequents" all points of intersection obtained from an intersection point fixed as initial (Fig. 1.7)

What can we expect from this process? Suppose that the system is integrable, that is, there are two conservation laws, the energy conservation,

$$E = E(x, y, u, v), \tag{1.26}$$

and a second conservation law,

$$F = F(x, y, u, v). \tag{1.27}$$

We can use the function Ω obtained by the algebraic solution of the energy integral to eliminate v from the latter conservation law. That law then becomes $F = \Phi(x, y, u; E)$, which is the conservation law that must be satisfied by the system reduced to 3 equations. But $F = \Phi(x, y, u; E)$ is the equation of a surface in the space x, y, u. Since it is a conservation law, every solution starting on the surface Φ must remain on it indefinitely. Then, when we cut the three-dimensional space with a plane and mark the points where the trajectory intersects this plane, the marked points must necessarily lie on the intersection of the surface

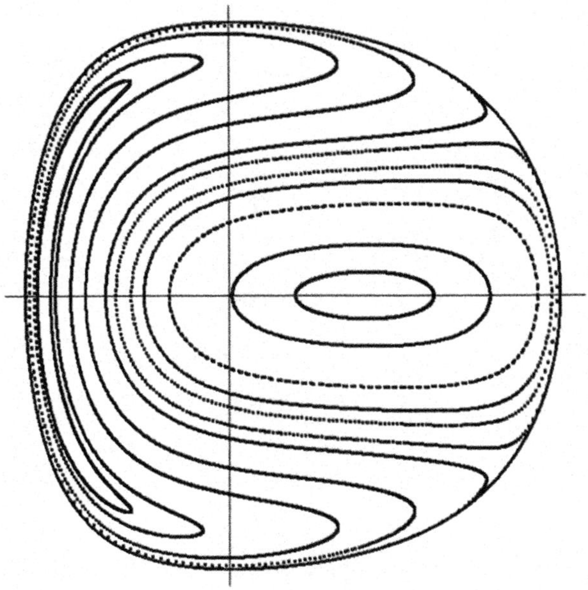

Fig. 1.8 Surface of Section (Poincaré map) of the Toda lattice, an integrable system with two degrees of freedom. The coordinates are one of the angles of the Toda lattice and its time derivative. The contour defines the domain where the solutions exist for a given energy.

$F = \Phi(x, y, u; E)$ with the plane, and nowhere else. Thus we conclude that when the integral $F = F(x, y, u, v)$ exists, the points marked on the Surface of Section must lie on a line (or on several segments of a line). Even if we do not know, in practice, how to solve the equations to obtain this second conservation law, a Poincaré map in which the consequents are arranged on lines is an indication that this second conservation law exists.

Figure 1.8 shows the Poincaré map (Surface of Section) of the Toda lattice. The Toda lattice is a mechanical system that seeks to represent a crystal lattice as simple as possible, consisting of three atoms arranged in a circle which repel one another with a force proportional to the exponential of the angle separating them: e^x, e^y, and e^z $(x + y + z = 2\pi)$. If one of the atoms is taken as a reference, the motion of the other two atoms relative to it is a mechanical system with two degrees of freedom; it is characterized, for instance, by the angles x and y between the atom taken as reference and the other two. The equations of motion are easy to obtain, and it is also easy to obtain the integral of the energy. Obtaining these equations is a simple exercise (Ford et al., 1973).

Figure 1.8 shows the consequents of a dozen different initial conditions, all with the same energy. In some of them, the number of iterations was large enough for the dots to form a line. In some segments, we can still see the dots, but always in perfect alignment. The ordered disposition of the consequents in the different curves indicates the existence of a second conservation law, in addition to energy conservation and, consequently, the integrability of Toda's lattice.

If, contrary to what is shown in Fig. 1.8, the marked points spread out in an irregular way, we have a sure indication that this second integral does not exist and, then, that the given system is not integrable (see examples in the next sections).

In practice, we use the four original equations and integrate them for a chosen set of initial conditions. Thus, for each of them, we obtain $x(t)$, $y(t)$, $u(t)$, $v(t)$ and, with the integration result, it is enough to determine the intersections of the solutions with a fixed plane crossed by the solutions to obtain a Surface of Section.

We could instead just integrate the three equations that resulted from using the energy integral to eliminate the variable v. However, numerical methods are not exact. So instead of eliminating the variable $v(t)$, as this does not affect the computational effort significantly, we integrate the entire system. In this case, the fourth variable may be used to verify the quality of the numerical solution by taking, at each instant, the values of x, y, u, v, and calculating $E = E(x, y, u, v)$. If the integrations are of good quality, the value of E obtained along a solution should remain constant.

Here comes an important technical detail. In the simplified reasoning done, we were intuitively using Euclidean geometry. But the usual systems are not so simple. Even if we assume that the phase space of the variables x, y, u, v is Euclidean, the hypersurface obtained by eliminating one of the variables needs not be Euclidean. It would be correct to say that the result of this operation, under a few simple conditions, is a differentiable manifold. For instance, if the equations we are studying refer to a mechanical system where the energy E introduces the square of the velocities, when we eliminate one of the variables using the energy integral, the resulting surface, in which the three remaining variables move, is not Euclidean. If the eliminated variable is, for instance, one of the velocity components, the function $\Omega(x, y, u; E)$ involves a square root. Therefore, for each value of x, y, u, we do not obtain one single value of v but two, one corresponding to the positive determination of the square root and another corresponding to its negative determination. That is, to each point of the reduced three-dimensional space that we obtained by restricting ourselves to a given value of the energy E, there correspond two values of the original 4-dimensional phase space. If we then mark over the plane used to cut the three-dimensional space, all intersections, indistinctly, we will be superimposing, on the same plane, the intersections of two distinct families of solutions. It follows that we should not mark all intersections with the fixed plane but only those having one chosen sign of v.

In mechanical problems, we use the integral of energy to eliminate one of the components of velocity, say $v = \dot{y}$, and we make the intersection using a plane $y =$ const. In this case, the identification is easy: it is enough to select only those points where the plane is crossed in one given direction. Obviously, this precaution is not necessary in those cases where the function $\Omega(x, y, u; E)$ has a unique determination.

As this is a method in which the diagnostics of integrability is made through a graphical operation, there are limitations. Once the graphs are made, if the points

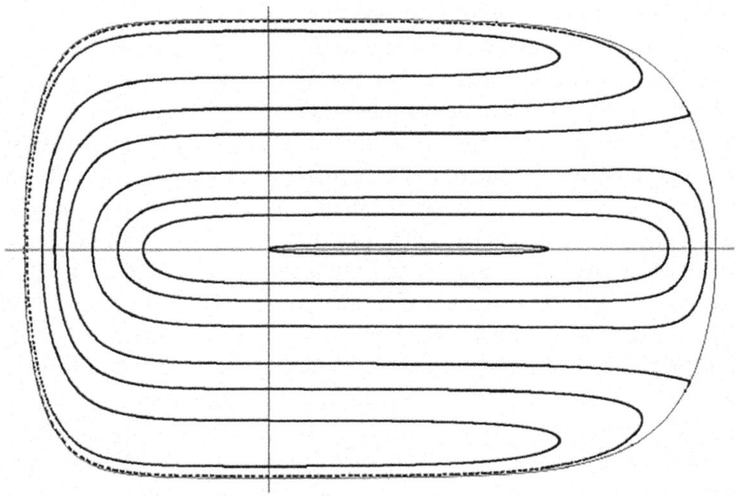

Fig. 1.9 Surface of Section of the Toda lattice for an energy 100 times larger than in Fig. 1.8

corresponding to each of the solutions appear to form a line or line segments, the system will be diagnosed as integrable. However, several things can happen that contradict such a conclusion. For instance, a magnification of the figure may show that the points are not really forming a line but are spread out (see Fig.1.11), forming a layer so thin that, at first sight, they appear to form a line. It can also happen that, when we continue the construction of the solutions for a longer time, the points that were appearing as if they were on a well-drawn curve begin, after a certain time, spread out visibly in their neighborhood.

A more common case is that in a Poincaré map, the lines corresponding to some initial conditions appear well-defined and remain so even if the integration time is increased, while for other initial conditions, the same system leads to points with visible scattering. In these cases, in fact, the system is non-integrable. They will be studied in Sect. 1.13.1. An important characteristic of integrable systems is that the integrability does not depend on the energy. In the system studied in Sect. 1.13.1, the apparent integrability of the system disappears when the energy is increased. In the case of the Toda lattice, the Surfaces of Section constructed for very high energies continue showing the same features as in low energies. Figure 1.9 shows the Surface of Section of the Toda lattice for an energy 100 times larger than in Fig. 1.8. The comparison of both figures shows that they are topologically equivalent. We may pass from one to another by a mere continuous deformation. And this is so even for much higher energies (Ford et al., 1973). The problem with the construction of the Surfaces of Section in high energies is that the solutions will cross the section with higher velocities deteriorating the quality of the calculation of the point where the section is crossed. If some precautions are not taken (such as a smaller integration step or better interpolation formulas), the curves may appear blurred due to the imprecise determination of the crossing point.

In some critical cases, it will be necessary to complement the Surfaces of Section with some of the other methods described in this book. Looking at the Fourier transform of some solutions may help to decide if they correspond to a regular or a chaotic motion. In Sect. 3.16, an example is given where various sections look alike, and their correct identification is done by constructing the dynamical power spectrum of the solutions. In this case, the merit of the Surfaces of Section is to show the topology of the invariant surfaces in the phase space, information that the other methods generally do not provide.

Every numerical diagnosis of integrability has a limitation. Finding a well-drawn line on a Surface of Section is not the same as finding the missing conservation law. In this case, once the conservation laws are found, the integrability of the system is determined. But with numerical diagnostics, there always remains the possibility that the degree of discernment with which we see the points in the graph may cause us to interpret a set of points that does not define a line as a line. Once more, we note that, as in previous paragraphs, we are not talking about inaccuracy of calculations. In this discussion, we are always assuming that the calculations are done with a perfect integrator, a perfect computer, and that they are, therefore, error-free.

Let us interpret the limiting cases discussed above. Suppose that the result looks like a line, but its enlargement shows that the points have an irregular distribution and do not form a line. This means, mathematically, that an additional conservation law does not exist in this system besides the conservation of energy. However, from a physical point of view, the quasi-alignment of the points can be interpreted as indicating the existence of an approximate conservation law. A law that is obeyed within a certain error. If this quantity $F = F(x, y, u, v)$, which corresponds to an approximate conservation law, is a quantity that can be measured, we will verify that all measurements along a solution will lead to the same value, within the intrinsic error of the measurements. Thus, even if a system is not integrable, if the numerical experiments show an indication of integrability, the system behaves as if it were integrable, at least for times of the order of the time of the simulation,

It is also necessary to consider the time limitation of the numerical integration performed. The numerically diagnosed integrability is only valid for a time of the order of the duration of the performed integration. In the study of an astronomical phenomenon, if it is necessary to integrate for a time longer than the age of the universe to prove non-integrability, we can consider that the system evolves as if it were integrable! Another example is that of a beam of ions in a Tokamak, whose duration is measured in milliseconds. If the time required to highlight the non-integrability of the beam equations is of seconds, then, in this case, also, we can consider the system as if it were integrable. In both examples, the time required for an irregularity due to non-integrability to manifest itself is much longer than the lifetime of the phenomenon being considered.

1.13 The Motion of a Star in a Galaxy with Axial Symmetry

Let us consider the Sun and the nearby stars. As the number of stars in the Galaxy is very large, it is impossible to consider an N-body problem that includes the mutual interaction of all of them. To study their dynamics, we use approaches based on statistical mechanics. We consider the phase space of 6 dimensions (3 degrees of freedom) $x, y, z, \dot{x}, \dot{y}, \dot{z}$. The distribution of the stars in this space of positions and velocities is defined by a distribution function $f(\mathbf{r}, \mathbf{v})$ which is equal to the probability of finding one star in the elementary volume defined by the opposite vertices $(\mathbf{r}, \mathbf{v})$ and $(\mathbf{r}+\mathrm{d}\mathbf{r}, \mathbf{v}+\mathrm{d}\mathbf{v})$. We assume that this distribution function f does not depend on time, that is, that the system is stationary.

If we know the distribution $f(\mathbf{r}, \mathbf{v})$ of the stars in the vicinity of the Sun, statistics gives rules that allow us to calculate the parameters of this particular distribution (Kurth, 1967). For instance, the dispersion of the components of the velocity in the direction of the Galaxy center ($\dot{\rho}$) and in the direction perpendicular to the galactic plane ($\dot{z}$) are calculated, respectively, by the integrals:

$$\sigma_{\dot{\rho}} = \int \dot{\rho}^2 f(\mathbf{r}, \mathbf{v}) \mathbf{drdv}, \tag{1.28}$$

$$\sigma_{\dot{z}} = \int \dot{z}^2 f(\mathbf{r}, \mathbf{v}) \mathbf{drdv}. \tag{1.29}$$

extending the integration over all the stars close to the Sun (In these approximated formulas, the average of the relative velocities is assumed equal to zero).

There is a theorem in classical statistical mechanics, due to Jeans (1961), which states the following:

> If the individual acceleration of the particles derives from a potential, that is, if the equations of motion are of the type $\ddot{\mathbf{r}} = -\mathrm{grad}\, U(\mathbf{r})$, then, necessarily, the distribution function $f(\mathbf{r}, \mathbf{v})$ must be a function of the first integrals of the motion.

The Sun and the nearby stars are inside a spiral galaxy, between its center and its edge. The galaxy rotates about an axis passing through its center and is reasonably symmetric about this axis, so that the force field created by it shows rotational symmetry. Then, the potential U created by the galaxy depends only on the distance ρ of the star to the rotation axis of the galaxy and the coordinate z. Because of the rotation symmetry, the field does not depend on the longitude θ. Hence, it follows that the equations of motion have two conservation laws: the conservation law of energy (in cylindrical coordinates): $\frac{1}{2}(\dot{\rho}^2 + \rho^2\dot{\theta}^2 + \dot{z}^2) + U(\rho, z) = E$, and the conservation law of the angular momentum projected onto the axis of rotation: $\rho^2\dot{\theta} = F = \mathrm{const}$.

Then, we only have two functions, E and F, to model what is happening, and the distribution function f only depends on E and F. Therefore, f will depend on the velocities $\dot{\rho}$ and $\dot{z}$ only through the sum $\dot{\rho}^2+\dot{z}^2$, which appears in the energy integral (the projected angular momentum conservation law does not depend on these two velocity components). Then, the two integrals that come from statistics and allow us to calculate the dispersions of the two components of the velocity are equal (If we exchange $\dot{\rho}$ for $\dot{z}$, one integral is converted into the other). That is, the dispersions of the velocities $\dot{\rho}$ and $\dot{z}$ must be equal.

This can be verified, as the velocity of the stars in the vicinity of the Sun can be measured by usual astronomical methods. Let us consider the radial velocities of the stars, that is their velocities projected on the line connecting the star to the Sun. For simplicity, we can first consider only the stars located in the direction of the center of the galaxy. In this case, the measured radial velocity is $\dot{\rho}$ (with an offset due to the motion of the Sun). As the largest component of the velocity of the stars is the one which makes them revolve around the center of the galaxy, the $\dot{\rho}$ components are smaller than the total velocities and may be positive or negative. We can draw a histogram of the measured velocities and determine their dispersion. We can then repeat the same procedure with $\dot{z}$. For this purpose, we consider the radial velocity of the stars that are in the direction of the galactic poles. Again, we make a histogram and determine the corresponding dispersion.

By the above theoretical result, the observed dispersions of the stars in the vicinity of the Sun, in the direction of the center of the Galaxy, and in the direction perpendicular to the plane of the Galaxy should be equal. But they are not! The observed dispersion of $\dot{\rho}$ is twice as large as the dispersion of $\dot{z}$. The observed result is very different from the calculated result. What is the cause of this error? The cause is that we only have two conservation laws to use in the distribution function. The disagreement between the predicted result when we assume that only these two conservation laws exist and the observed result can be considered an important indication of the existence of one more conservation law.

1.13.1 The Hénon-Heiles System

The problem of the difference in the dispersions of the stars' velocities in the solar neighborhood, in the directions of ρ and z, was intensively studied in the 1960s by several astronomers, in particular, George Contopoulos, who investigated the nature of a possible third integral (Contopoulos, 2002). The solution to the problem came from works based on the ideas of Poincaré (1899). The motion of a particle in a galaxy with rotation symmetry is initially a problem with 3 degrees of freedom given by 6 first-order differential equations. But the variable θ (longitude) does not appear on the right-hand sides of these equations. So, we do not need to consider it. Neither do we need to deal with $\dot{\theta}$, because, knowing ρ and the angular momentum, the conservation law of angular momentum gives $\dot{\theta} = F/\rho^2$ allowing us to know it. We are left with only the equations in ρ, z, $\dot{\rho}$, and $\dot{z}$, that is, a system with two degrees of freedom that describes the motion of the star in a meridian plane (which contains the axis of the galaxy) that rotates following the star in its orbital motion.

The resulting system has two degrees of freedom and obeys the law of conservation of energy. To study it, Hénon and Heiles (1964) have done some simplifications, changed the variables from ρ, z to x, y and introduced the potential

$$U(x, y) = \frac{1}{2}(x^2 + y^2 + 2x^2y - \frac{2}{3}y^3). \tag{1.30}$$

Then, they studied the resulting equations:

$$\begin{aligned} \ddot{x} &= -x(1 + 2y) \\ \ddot{y} &= -y - x^2 + y^2 \end{aligned} \tag{1.31}$$

using the method of Surfaces of Section.

This study was the first application of this method to a specific problem since its proposal by Poincaré 65 years earlier. The method of the Surfaces of Section, as we have already seen, is adequate to answer the question of whether there is a second law of conservation in this system besides the energy conservation. That is, a third conservation law in the complete system (also including the projected angular momentum conservation law used to reduce the given system to two degrees of freedom). If the complete system has a third conservation law, then this reduced system must have another conservation law besides the energy conservation law.

Let's look at the results obtained. The first panel of Fig. 1.10 corresponds to a low energy value. Note that the curves obtained are closed as in the case where there is another additional conservation law and the system is integrable. The ovals are intersections of bundles of concentric tubes, which close after one turn. In the center of each of them lie stable periodic orbits. There are some saddle points, and the branches originating at one of them end in the other (red lines). As it happens in an integrable system! The saddle points are intersections of unstable periodic orbits. A solution that moves away from it a little moves away from it forever.

The second panel of Fig. 1.10 corresponds to slightly higher energy. The four centers of the previous figure still exist. But when we move away from the center of these islands and approach where the corresponding saddle points should be, the lines break up: the consequents start jumping from one point to another in an apparently random way. In the previous panel, the points were distributed in such a way as if it was always possible to pass a line over them. Now, it is no longer possible to do so. The regular regions that appeared in the previous panel, near the points of stable equilibrium, have been preserved, but near the branches connecting the saddle points, which in the previous figure appeared to separate different oscillatory regimes, continuous smooth lines no longer exist. All points appearing in this region are consequents of a single solution. And this stochastic region is extensive. This shows that a new conservation law does not exist and the Hénon-Heiles system is non-integrable.

To complete the picture, we include a third panel for even higher energy, twice the energy of the first panel. In this figure, the regular regions practically disappeared.

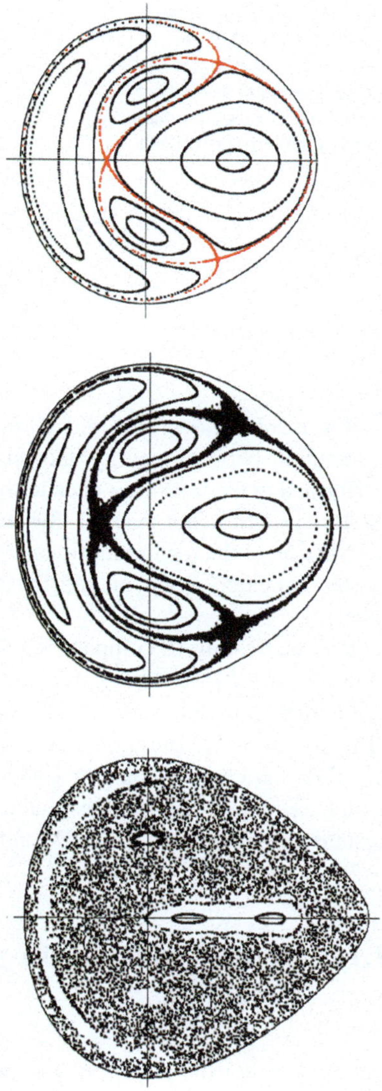

Fig. 1.10 Surfaces of Section of the Hénon-Heiles system for various energies (from top to bottom, 0.08, 0.10 and 0.16). The contours define the domain where the solutions exist for the given energies. The coordinates are the variables y, $\dot{y}$ (see Eq. 1.31) and the section is defined by the plane $x = 0$ with the condition $\dot{x} > 0$

All the scattered points appearing on this Surface of Section are consequents of a single initial condition. They belong to the same solution.

In the first panel, we do not "see" any stochastic regions. But, in fact, they are there. They are not visible because of the always limited approximation used when making a figure. But if we enlarge this figure, we see that there are also stochastic regions there. But as these are in very thin layers, they disappear in the imprecision of the drawing. In particular, if we zoomed in on the figure near the leftmost unstable points, what looked like a saddle point with characteristics equal to those of an integrable system appear quite differently as shown in Fig. 1.11.

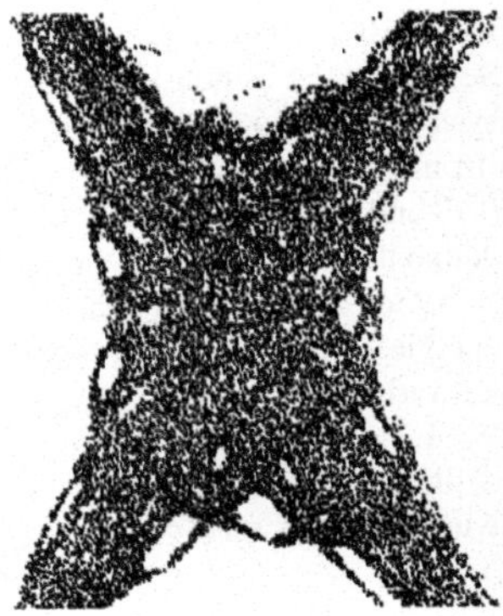

Fig. 1.11 Magnification of the neighborhood of the unstable equilibrium located on the horizontal axis where the red lines of the first map of Fig. 1.10 (top) cross each other

The results shown in Fig. 1.10 may give the impression that there might be an association between energy and the disappearance of regular motions. But work performed in the following decade showed that this fact, verified with the Hénon and Heiles system, is not general. Systems were found in which, after the almost total destruction of the regular motions, by increasing the energy even more, well-defined curves appeared again on the Surface of Section in large numbers (Yokoyama, 1984).

It is interesting to compare the above results with those obtained for the Toda lattice (Fig. 1.8), in which the consequents are always located on curves, never appearing scattered, not even in the vicinity of the saddle points. They are quite different from the previous ones. However, the development of the equations of the Toda lattice gives, as a first approximation, the equations of the Hénon-Heiles system. Higher-order infinitesimal terms dramatically change the dynamics of that system (see Ferraz-Mello, 2007) and result in its integrability.

1.14 Systems with One Degree of Freedom. The Simple Pendulum

Let us consider the phase plane of a mechanical system with one single degree of freedom. Because of the incompressibility of the flow, there cannot exist in this plane any sources or sinks of trajectories. The only possible fixed points are centers and saddle points, or compositions of them formed by the occasional coalescence of two or more centers and saddle points. The centers correspond to points of stable equilibrium, and the saddle points to points of unstable equilibrium. From the unstable equilibrium points four branches always emerge; two stable and two unstable. The stable branches are paths along which the motion tends asymptotically to the equilibrium point, reaching it in an infinite time. The unstable branches are paths along which the motion has the opposite direction to that of the stable branches, moving away from the equilibrium point that they left at an infinite time in

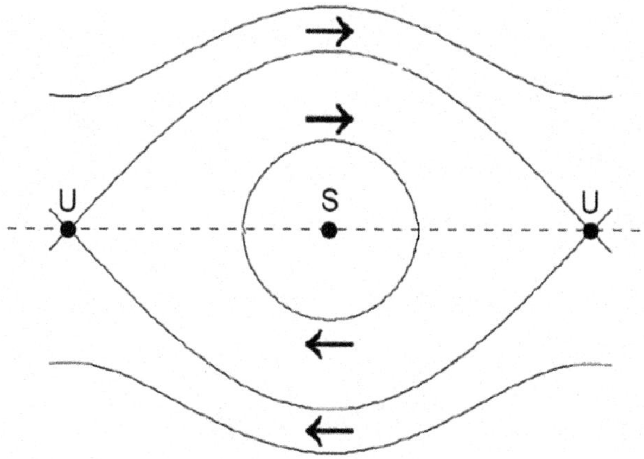

Fig. 1.12 Phase portrait of the simple pendulum. Abscissas: angle of position θ; ordinates: angular velocity $\dot{\theta}$; dashed line: $\dot{\theta} = 0$. Equilibrium points: center S: stable ($\theta = 0$); saddle U: unstable ($\theta = -\pi$ and $\theta = +\pi$)

the past. In the two stable branches, the solution tends to equilibrium when $t \to \infty$ and, in the two unstable branches, when $t \to -\infty$.

Let us take as an example a pendulum with its two points of equilibrium: the stable equilibrium point at the bottom, and the unstable equilibrium point, at the top. In the phase plane $(\theta, \dot{\theta})$, we have (Fig. 1.12) a center which is the stable equilibrium point (S). When we move the pendulum a little away from (S), it will oscillate. In the phase plane, the oscillation around the stable equilibrium is represented by a closed curve. $\dot{\theta}$ is positive in the upper half-plane and negative in the lower half-plane. By increasing the amplitude, we have curves that are larger and involve the previous ones. Increasing the initial velocity even more, we reach a value at which the pendulum can reach the highest point of the motion; in this motion, the velocity decreases as the point moves away from the stable equilibrium and approaches the highest point. If the energy is exactly that required to reach the highest point, the time to reach it is infinite, and the motion is such that the velocity of the pendulum tends to zero as the pendulum approaches the unstable equilibrium position (U). The trajectories reaching the unstable equilibrium point by both sides are the so-called stable branches; the trajectories leaving the unstable equilibrium point, also by both sides, are the so-called unstable branches (Fig. 1.12).

Let us recall that, in the pendulum, a motion which is asymptotically tending to the unstable equilibrium point on one side, in the future, tends asymptotically, in the past, to the same point, but from the other side. The stable and unstable asymptotic branches meet at the unstable equilibrium point two by two. Outside the region delimited by the asymptotic branches coming out of the saddle points, the trajectories correspond to a pendulum with enough energy to keep itself turning (circulating).

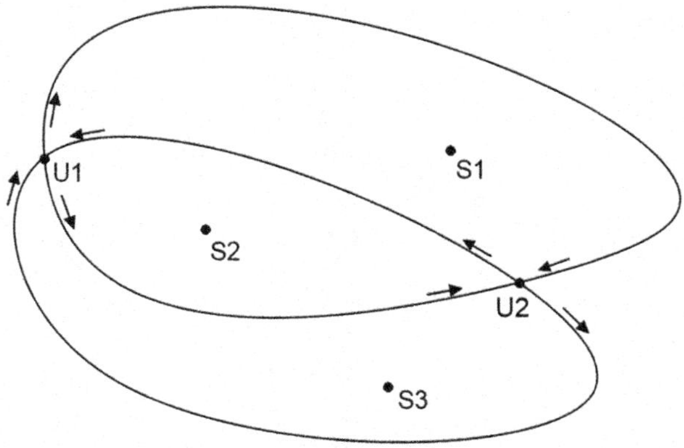

Fig. 1.13 Phase portrait of a one-degree-of-freedom system with two saddle points and three centers

Let us now think of a generic problem with two unstable equilibrium points and consider two asymptotic paths, each one exiting one of the unstable equilibrium points (saddles). Can they intersect? The answer is no. Just remember that if the differential equations defining the system have continuous and differentiable right-hand sides, the Lipshitz theorem says that only one solution can pass through each point. Then, either the point is a singularity—as occurs in unstable equilibrium points—or no two solutions can pass through it. If two such curves were allowed to cross, two solutions would be passing through the same point. Thus, if a stable branch arriving at a saddle point meets an unstable branch leaving the same or another saddle point, the angle formed by the two branches at the meeting point must be equal to zero so that one is the differentiable continuation of the other.

Asymptotic branches that meet merge into a single trajectory which is stable for one saddle point and unstable for the other. An asymptotic branch that does not go to infinity is never simply stable or unstable. It is unstable with respect to the past (it is leaving a saddle point) and stable with respect to the future (it is entering a saddle point).

If the phase plane has two saddles, there are 8 asymptotic branches coming out of them (Fig. 1.13). If none of them goes to infinity, these branches will have to meet 2 by 2. The asymptotic branches define domains, each enclosing in its interior one stable equilibrium point (in the figure, the points S_1, S_2, S_3).

1.15 Two Degrees of Freedom

We now turn to systems with two degrees of freedom and consider first their periodic solutions. In a Surface of Section, a periodic orbit is marked by a fixed point. At each period, the orbit returns to the same point. If the orbit crosses the Surface of Section several times until it completes one period, we have a family with a finite number of points. In that case, each of them is a fixed point of the transformation that corresponds to solutions crossing the Surface of Section that same number of times.

The incompressibility property of the flow is preserved by the Surface of Section construction operation and must be verified by the mapping of a set of points into their consequents. Therefore, the fixed points of the Surface of Section can only be centers or saddle points (see Fig. 1.10).[6] The centers correspond to stable periodic orbits: A solution that starts from an initial condition close to a center remains forever near the center. The saddle points are intersections of unstable periodic orbits. The asymptotically stable and unstable branches emanating from a saddle point on the Surface of Section now mark intersections of surfaces on which are located trajectories that are asymptotically going away from or approaching the unstable periodic orbit. These surfaces are called stable manifolds and unstable manifolds, respectively.

What do the asymptotic branches bifurcations look like in this case? This question is crucial. Will the branches coming out of the unstable fixed points of the Surface of Section necessarily close, as in the case of one degree of freedom? The argument used to show, in that case, that stable and unstable asymptotic trajectories never cross was the uniqueness of the solutions. Here, the uniqueness must also be respected. But it does not have the same implications.

Let us take two branches that come out of the same or of two distinct saddle points and reach a point A, where they intersect. Suppose that the meeting point is not another saddle point and that they cross at that point, forming a non-zero angle (Fig. 1.14).

The points where a stable branch and an unstable branch meet, both originating from the same saddle point, are called homoclinic points. On the Surface of Section, the stable and unstable branches are no longer trajectories but lines marking the intersections with the Surface of Section of trajectories that lie on manifolds formed by solutions asymptotic to an unstable periodic solution. We know the following: If a trajectory is on a stable (or unstable) manifold, each time it intersects the Surface of Section, it does so at a point that is on the same stable (or unstable) branch approaching (or respectively, coming out of) a saddle point. Then, if two manifolds, one stable and one unstable, intersect, there are two branches, one stable and one unstable, meeting at one point in the Surface of Section. The solution that passes through the meeting point belongs to both manifolds. Its consequent, that is, the point at which it will cross the Surface of Section again, must also belong to the same two manifolds: that is, it must be a new crossing point of the two branches. This new crossing point cannot be the same point from which the branch started,

[6] In the Surfaces of Section of dissipative systems, the flow may contract and sinks to which converge a bundle of solutions may exist. Sinks and other attractors are singularities ruled out in conservative systems. The divergence of the flow in their neighborhoods is not equal to zero (see Sect. 1.8).

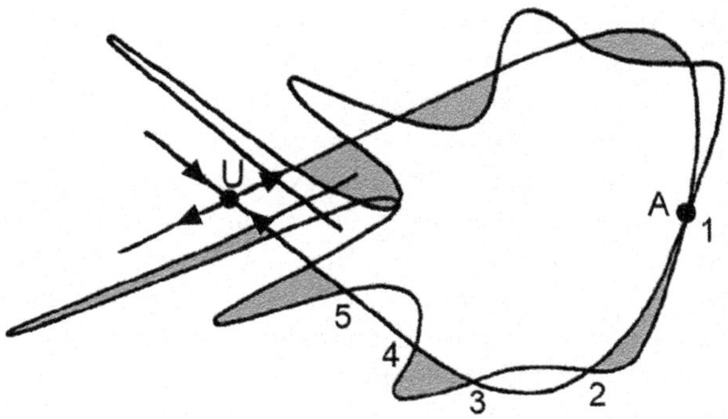

Fig. 1.14 Homoclinic points formed by the intersections of two asymptotic branches coming out of the same saddle point U

otherwise, this point should be a fixed point on the map, but we explicitly ruled out this possibility.

Thus, if two branches cross, they must do so again, and then again, etc., infinitely many times, both for the future and for the past. Let us mark with $1, 2, 3, 4, 5, \cdots$ all homoclinic points where the two branches cross again (in the future). We can also mark with other symbols the sequence of homoclinic points where the two branches may have crossed in the past. If we consider the dihedral formed by small segments coming from 1 and 2 along the two intersecting asymptotic branches, and oriented according to the orientation of these branches, the dihedrals formed with vertices at 1 and 2 have reversed orientations. Therefore, a bundle of trajectories cannot lead from the neighborhood of 1 into the neighborhood of 2, and 2 cannot be an image (consequent) of 1. The images of 1 must be $3, 5, \cdots$, points where the two branches intersect the Surface of Section with the same orientation as in 1.

The sequence of consequents $1, 2, 3, 4, 5, \cdots$ tend asymptotically to the saddle point where ends the branch on which they are located. Therefore, these points tend to be closer and closer to each other. On the other hand, we know that the transformation used in the construction of the Surface of Section, which leads from a point to its consequent, defines an incompressible flow. Therefore, all solutions leaving a given domain, when they intersect with the Surface of Section again, must form a domain of the same area as the original one. Thus, the areas enclosed by the corresponding loops must be equal (e.g., the darkened areas in Fig. 1.14 must be equal).

As the distances between the crossing points decrease tending asymptotically to zero, the loops tend to stretch out and have increasing lengths. As an additional complicating factor, let us remember that a branch cannot cross itself since, in such case, the uniqueness of the solutions would be violated. These facts make it extremely difficult to draw what occurs.

1.16 The Homoclinic Entanglement. Regime Transitions

The difficulty of graphically showing what happens with intersecting asymptotic branches is so big that in "Les Méthodes Nouvelles de la Mécanique Céleste", the book in which they were described for the first time, Henri Poincaré, its author, apologized for even not daring to try to draw them. Nowadays, with the help of computers, it is possible to do so, and the following figures are due to George Contopoulos and Charalampos Polymilis.

Figure 1.15 shows asymptotic branches emanating from the same saddle point and crossing each other two by two. The first figure shows the two stable branches and the two unstable branches existing in the map considered. In the second figure, only one unstable branch is shown, but for a time more than 3 times longer.

The entanglement of stable and unstable branches emanating from the same saddle point is called the homoclinic entanglement. It is called a heteroclinic entanglement if the branches emanate from two different saddle points. Following usage, the immediate neighborhood of the chaotic entanglement (homoclinic or heteroclinic) is called the stochastic layer. In fact, since the other solutions (outside the stable and unstable manifolds) cannot intersect, any solution close to one of these manifolds will have to follow it closely all along its tortuous path. The existence of homoclinic (or heteroclinic) intersections is an essential condition for the existence of chaotic solutions.

The Surfaces of Section of Sect. 1.13.1 show, in the vicinity of the saddle points, a crowded set of points. The corresponding solutions do not lie on a smooth surface.

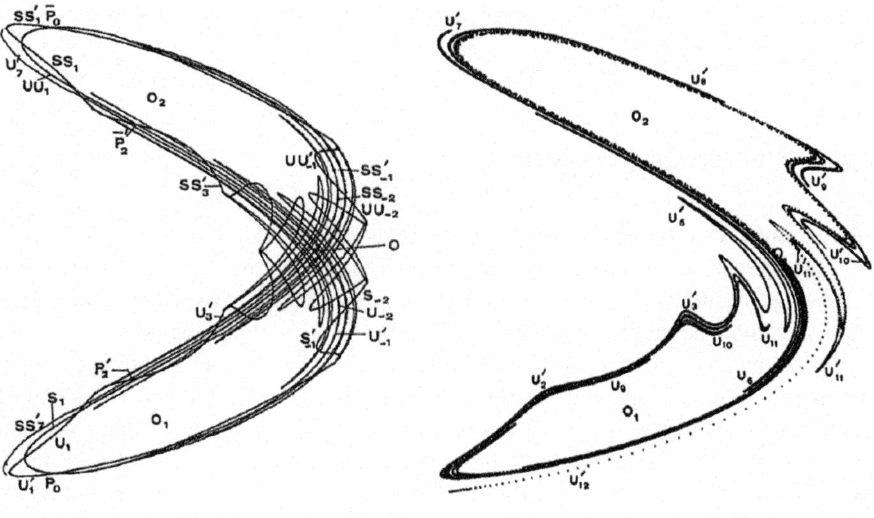

Fig. 1.15 Asymptotic branches from the same saddle point O. Left: figure complete. Right: one of the unstable branches extended for a longer time. (reprinted with permission from Contopoulos and Polymilis (1993). Copyright 1993 by the American Physical Society)

As we discussed previously, in these cases, there are no new conservation laws, and the system is not integrable.

When a two-degree of freedom system is an integrable system, the entanglement observed in the Surfaces of Section in the previous case cannot occur. In the case of an integrable system, the stable and unstable branches cannot intersect at angles other than zero. The figure must be like the one discussed in the case of a system with a single degree of freedom where, in general, stable and unstable branches coalesce into a single branch which is unstable with respect to a periodic solution to which it was close in the past infinity, and stable with respect to the one it will approach in the future infinity. Therefore, a new way to define non-integrable and integrable systems is, respectively, the existence or non-existence of homoclinic (or heteroclinic) points. Systems with a single degree of freedom (like the pendulum) are integrable systems: in a system with one single degree of freedom, there cannot be homoclinic (or heteroclinic) points, so chaotic solutions cannot exist.

To summarize, in an integrable system, the stable and unstable manifolds separate the phase space of the system into domains foliated by invariant surfaces. A motion originating at one of these surfaces remains there forever. Its regime of motion is immutable. For instance, if a perfect pendulum receives enough energy to rotate, it remains indefinitely in this state; if the initial energy is not enough to rotate, it will oscillate indefinitely. There are many examples in which this happens. In non-integrable systems, on the contrary, the stable and unstable manifolds are entangled in such a way that motions are allowed to transit from one regime of motion to another.

In the system shown in Fig. 1.15, solutions may exist that remain for a long time in a regime of motion characterized by oscillations around O_1 and then cross the boundary between the two domains and move to another regime of motion in which they oscillate around O_2.

1.17 Phobos and Hyperion

One example of a system in which two regimes of motion are mixed is the rotation of a satellite around a planet in the simple model in which the satellite is an ellipsoid revolving around one of its principal axes, perpendicular to the plane of orbital motion. Suppose that the equator of this satellite is very elongated, and let θ be the angle that gives the distance from the longest of the equatorial axes of the satellite to the line joining the centers of the planet and the satellite. The first panel of Fig. 1.16 corresponds to Phobos, the first Mars satellite. It shows that there are stable periodic solutions at $\theta = 0$ and $\theta = \pi$, which correspond to having the satellite rotation synchronous with its orbital motion. In this case, the satellite is moving around the planet in such a way that the longest of its equatorial axes always remains pointed toward the planet (like the Moon, which always shows the same face toward the Earth). Around each of these stable solutions, the satellite can have "librations". In this case, the face that the satellite shows to the planet oscillates, so that, in one

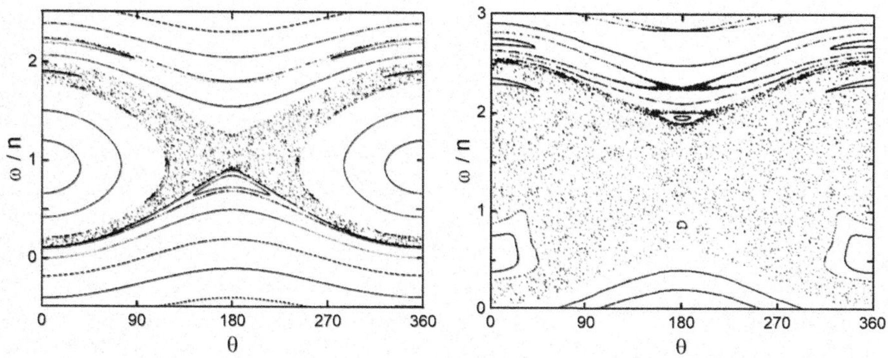

Fig. 1.16 Surfaces of Section in the phase space of the Phobos (left) and Hyperion (right) rotations ($\omega = \mathrm{d}\theta/\mathrm{d}t$) (Reprinted with permission from Wisdom (1987). Copyright 1987 by Elsevier)

period of oscillation, an observer in the planet can see more than 50% of the surface of the satellite. There are also solutions (in the upper and lower parts of the figure) in which the angle θ circulates. In this case, in a period of circulation of θ, all parts of the satellite can be seen from the planet. This figure shows even more. There is also an unstable solution at $\theta = \pi/2$ (in the middle of the figure) embedded in a chaotic zone that separates the two regimes of motion: libration and circulation. A motion that begins as a libration at the edge of this chaotic zone may turn into circulation after some time (or vice versa).

The second panel of the figure refers to Hyperion, Saturn's satellite, whose chaoticity has been considered before (see Sect. 1.4) and shows that the two regimes of motion, which appeared in the case of Phobos, simply disappear due to the strong chaos visible on the Surface of Section. We see only a few small islands, which should correspond to some periodic solutions, and a region of high-speed circulation, at the top of the figure (another, corresponding to rotation in the other direction, should occur when $\dot{\theta} < 0$, not shown in the figure). The difference in the two figures is due to the intensity of the force with which the planet disturbs the satellite rotation, which is proportional to the eccentricity of the satellite's orbit around the planet and to the ratio between the principal moments of inertia of the satellite. The latter, in the case of Hyperion, is quite large because of the large asymmetry of its figure ($190 \times 145 \times 115$ km).

In the case of the motion of the Moon around the Earth, one must have a figure similar to that of Phobos, but still with a much narrower chaotic layer because the Moon has a figure very close to a sphere; the actual motion of the Moon is an almost imperceptible oscillation around the stable periodic solution.

1.18 Enceladus and Dione

Another example is the problem of two objects orbiting a much larger body in orbits with commensurable periods. Several of these examples are studied in the chapter on the motion of resonant asteroids (Chap. 2). The example presented here refers to one of the first studies of chaos in orbits of this type using numerical integrations, and considers a pair of satellites of Saturn, Enceladus and Dione, whose orbital periods are close to the 2/1 commensurability (Ferraz-Mello and Dvorak, 1987). Indeed, because of the dissipation due to tidal effects, these satellites are very close to a stable periodic motion, and we have to exaggerate some parameters of the problem to see the whole dynamics of the system. It is a good example, but unlike the other cases shown in the following chapters of this book, it is too far from the real conditions.

In the obtained Surface of Section, there is a saddle point from which asymptotic branches emanate and entangle, creating the chaotic region shown in Fig. 1.17. All

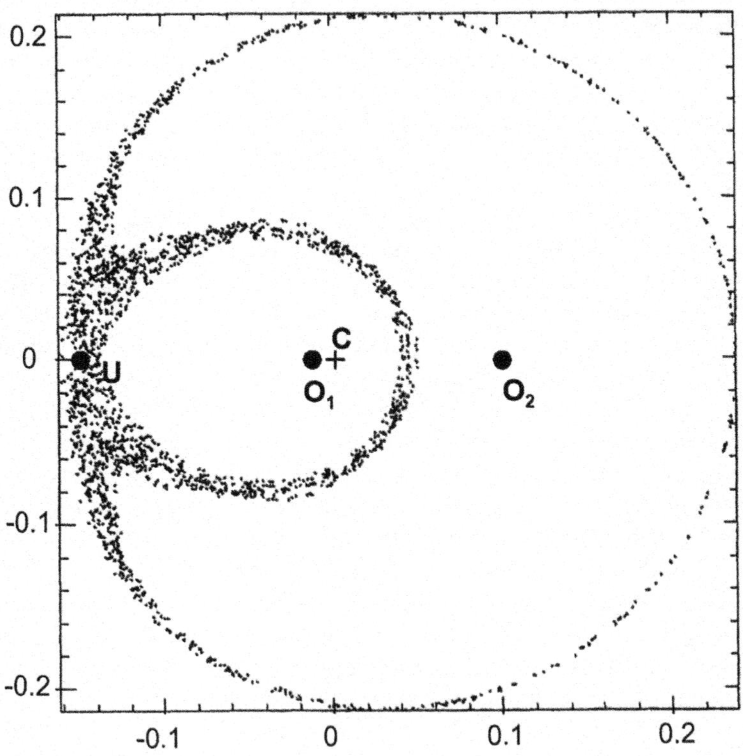

Fig. 1.17 Homoclinic entanglement in the Enceladus and Dione system. Polar coordinates with center at C. The radius vector is the eccentricity of the orbit of the inner satellite (maximum amplitude $\simeq 0.25$), and the angle is the resonant combination of the longitudes of the two satellites ($\psi = 2\lambda_D - \lambda_E$). O_1, O_2: centers (stable periodic orbits); U: saddle (unstable periodic orbit) (Reprinted with permission from Ferraz-Mello and Dvorak, 1987. ©ESO)

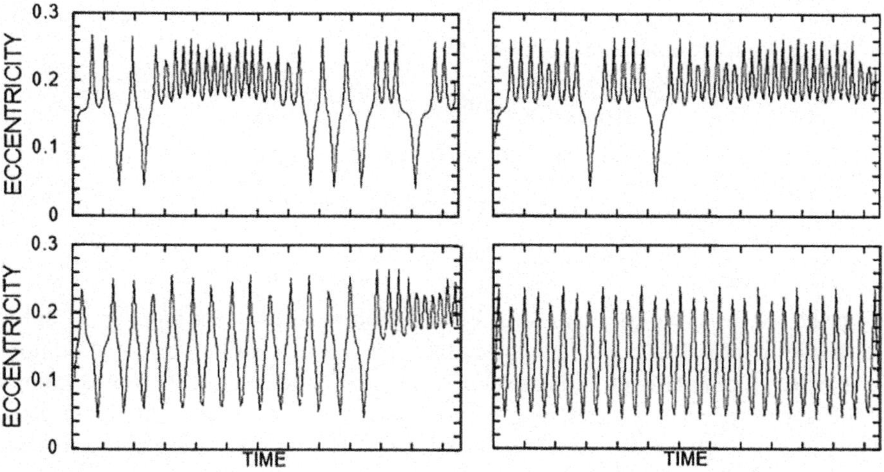

Fig. 1.18 Variation of the eccentricity of the inner satellite in apr. 600 years in solutions showing intermittencies between librations and external circulations (Reprinted with permission from Ferraz-Mello and Dvorak (1987). ©ESO)

the points shown are intersections with the Surface of Section of a single orbit over a long time span. The unstable orbit represented by the saddle point U generates the entire chaotic region.

By making small changes in the problem, namely small variations in the energy of the system, we obtained several types of solutions. In this system, we have three possible oscillation modes: (i) Internal circulations, which are motions that follow ovals around the center O_1; (ii) librations, which are motions that follow crescent shaped curves around the center O_2 (they are called librations because, in these motions, the polar angle oscillates without completing a revolution); and (iii) external, circumference-like circulations outside the chaotic zone, involving the two centers (Fig. 1.17). The real Enceladus motion is a quite small circle around O_1.

Figure 1.18 shows several solutions obtained. In three of these solutions, transitions between librations and external circulations occur. In the fourth solution, only external circulations appear.

1.18.1 Intermittencies and Symbolic Mechanics

The solutions shown in Fig. 1.18, where two motion regimes alternate, suggest that the intermittencies can follow alternation rules like those of the systems studied in symbolic mechanics. For instance, we can represent these solutions by successions formed only by the numbers 1 and 0. The number 0 is used to represent librations, and the number 1 to represent external circulations. Thus, the solutions shown in Fig. 1.18 are represented, respectively, by

1011000000000000000000011100100,

10000000010000010000000000000000000000000,

11111111111111100000000000,

1111111111111111111111111111111.

In principle, by arbitrarily choosing a succession formed by the numbers 0 and 1, it should be possible to find a solution that reproduces it in the form of librations and external circulations, respectively. For any given sequence, there is an initial condition in which the alternation of oscillations of one mode and another gives that sequence. This is not proved in problems of this nature, where the results are obtained by numerical integration of differential equations, but in simpler dynamical systems such as the logistical map (see Guckenheimer and Holmes, 1983 and Lichtenberg and Lieberman, 1983). In the solution of an equation it is always possible to do something like the following: We want a certain sequence to happen. For instance, let it be the first of the above sequences. We take the first number of this sequence (the number 1) and look for all initial conditions that lead to the occurrence of the event labeled with the number 1 (in our example, an external circulation). The set of initial conditions that lead to this event has a certain size. If we examine this set, there is a part of it where the next event is number 1 and another part where it is number 0. Since the second number in the chosen sequence is 0, we restrict ourselves to the part that corresponds to 0 as the second event. Again, the set of initial conditions selected has a certain size, and within it, one part has the event number 1 as third event, and another has the event number 0. Since the third number of the sequence we have chosen is 1, we restrict ourselves to the part that leads to 1 as the third event. And so on, indefinitely.

1.19 Perturbed Systems. Resonance and Libration

Let us consider the Surface of Section of an integrable system with two degrees of freedom whose phase flow is incompressible. Let us suppose it to be as simple as possible, with a stable fixed point, a center, around which successive closed circles exist, one after the other (Fig. 1.19a). The existence of these circles indicates the existence of a conservation law, and different circles correspond to different values of the conserved quantity. It is easy to visualize that, in this case, the surfaces intersecting the Surface of Section in closed circles are tori. Each torus corresponds to a different value of the second integral. A solution that starts on one torus remains on it indefinitely. These tori are, therefore, invariant surfaces. A simple point to check is that two invariant curves (i.e., intersections of the invariant surfaces with the Surface of Section) cannot intersect at a point that is not a fixed point (otherwise, there is a violation of the uniqueness of the solution at that point).

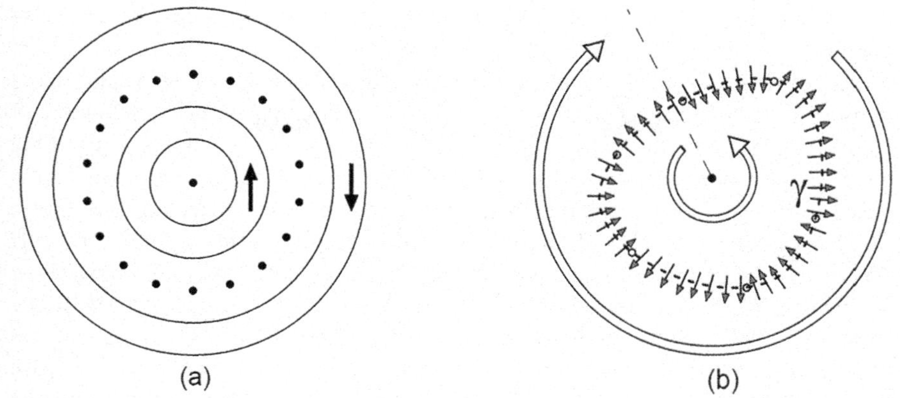

Fig. 1.19 **(a)**: Surface of Section of a simple integrable system with an inversion of the direction of rotation around the center (resonance); **(b)**: Direction of the radial displacement of the resonance line under the effect of a perturbation

Let us now suppose something else. Suppose that, in the innermost invariant curves, the successive intersections (or, in Poincaré's language, the consequents) circulate the curve in a given direction (counterclockwise in the example shown), while in the outermost invariant curves, they circulate it in the opposite direction (clockwise). As the frequency of this circulation must vary with continuity, in some curves, it cancels out. In this particular invariant curve on which the frequency is zero, the points do not move. That is, the solutions that intersect the Surface of Section along this curve are periodic. They are solutions that return to the same point at each intersection with the Surface of Section; they are periodic solutions of period 1 as they return to the starting point after one revolution. In mechanics, it is said that resonance occurs on this curve.

We want to know what happens to this integrable system when we add a perturbation to the right-hand side of the equations. This perturbation is, by hypothesis, quite small, so that in the innermost parts of the section, the solutions keep moving to one side, and in the parts farther from the center, to the other side. Let us consider a simplified version of the theories of Henri Poincaré and George David Birkhoff and the outlines of their main result.

Let us consider a family of rays going from the center to the parts away from the center. At one end of the ray, the points circulate in one direction, and at the other end, they circulate in the opposite direction. Due to continuity, at some point in the ray, there must be an inversion and, therefore, a point where the displacement transverse to the ray (that is, the circulation) is zero. On each ray, there is one point where a change of direction occurs. If the family of rays is continuous, the points where the change occurs define a line separating solutions circulating in opposite directions. It is evident that on each ray, there may be more than one point at which this occurs, and the set of points where the direction of the transverse displacement is reversed is not necessarily connex. The fact that matters is the existence of a closed line going around the center, on which the displacement transverse to the rays is zero. We call it γ. If there is more than one, everything we say about γ also applies to the others.

The above reasoning was limited to the vanishing on each ray of the displacement transverse to it, defining a curve γ. It remains to consider the component of displacement along the rays, that is, radial. At first, we cannot say anything about these radial displacements, and along the line, there are parts where the radial displacement is oriented outward and parts where it is oriented inward (see Fig. 1.19b). Question: Can all points have radial displacement oriented outward? The answer is no. We cannot have all radial displacements oriented outward (or all inward) because, in that case, at the next intersection, the line formed by the solutions initiated on the curve γ would form a new curve γ' enclosing an area larger (resp. smaller) than the area enclosed by the curve γ. This cannot occur because the hypothesis of incompressibility of the flow means that the areas enclosed by the curves γ and γ' must be equal. Therefore, the two curves must intersect one another. Then, either the radial displacements over the points of the curve γ are all zero, or they are distributed in parts where the radial displacement is turned inward and parts where it is turned outward.

Excluding exceptional situations, the radial displacement will be zero at an even number of points[7]. In these points, both transverse and radial displacements are zero: that is, these points are fixed points of the system, points where there is no displacement. After one revolution, the trajectory will return to its starting point. They correspond to periodic solutions. On the line γ, these points are alternately centers and saddle points, that is, crossings of stable and unstable periodic orbits, respectively (Fig. 1.20). We can identify them simply by looking at the direction of the transverse and radial displacements around them in Fig. 1.19b.

Two stable and two unstable branches come out from each saddle point. If the perturbation does not destroy the integrability of the system, these asymptotic branches do not cross one another. They meet in such a way that the stable branches appear as continuations of unstable branches. The displacements on them tend to a saddle point in the past and also in the future. These branches enclose zones where motion occurs around a center. Astronomers usually call "librations" the motion in the zones thus created by the perturbation. Each libration zone corresponds to a distinct regime of motion. Hence, in the resulting system, we have the internal and external zones, in which "circulations" occur in both directions and, between them, a region in which one or more libration zones are formed. They are separated by asymptotic branches, which for this reason, are usually called "separatrices" or "bifurcations".

If the introduction of the perturbation in the integrable system transforms it into a non-integrable system, then in the area enclosing the saddle points appear a domain where the asymptotic branches intertwine to form a stochastic layer in the Surface of Section.

[7] The exceptional cases excluded are, for example, those where the curves γ and γ' are tangent at one point. They are possible, but the ordinary case with an even number of intersections is sufficient for what is discussed in this section.

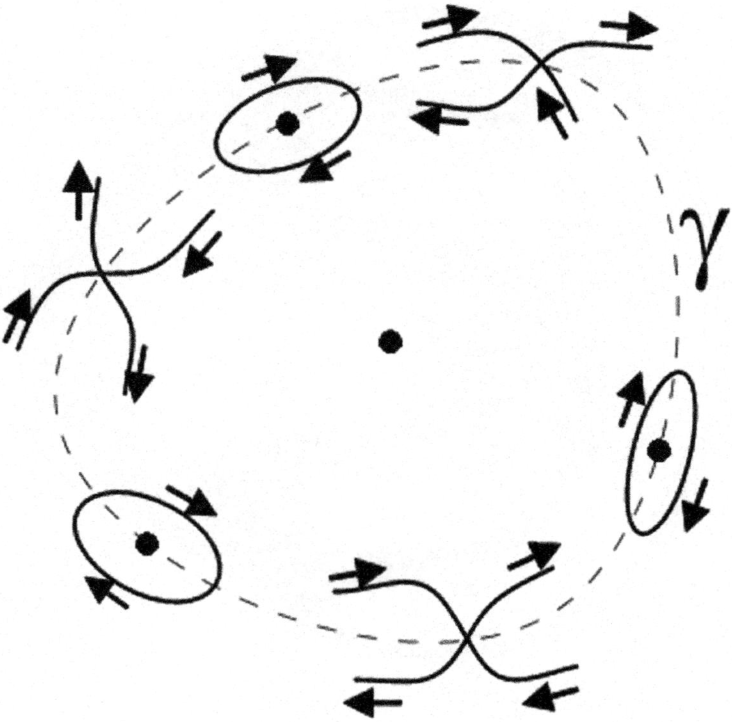

Fig. 1.20 Formation of centers and saddle points on the Surface of Section of the perturbed system along line γ

We have discussed possible behaviors of the solutions of an integrable system when a small perturbation is added to it. We showed that, due to the change of directions of circulation of the consequents in the Surface of Section, new situations may appear: libration zones if the perturbed system remains integrable, or chaotic entanglement if the perturbation destroys the integrability of the system. The word "may" was used for obvious reasons. It has not been proved anywhere that this must occur! For instance, the perturbation may result in only some deformation of the curves of the Surface of Section of the given integrable system, with no qualitative change in its solutions.

The situation above described for periodic solutions of period 1 is analogous for all other periodic solutions of the system. Consider, for instance, the periodic orbits of period N, i.e., periodic orbits that close after crossing the Surface of Section N times. If we look at the invariant curves outside and inside the curve where these periodic orbits are located, we see a situation analogous to the previous one. If we consider the transformation that consists of marking the Nth intersections while omitting the others (let us call it T^N), we may repeat for it all operations done in the more simple case studied before: In the place where the periodic orbits of period N were originally located, new centers and saddle points appear, and from

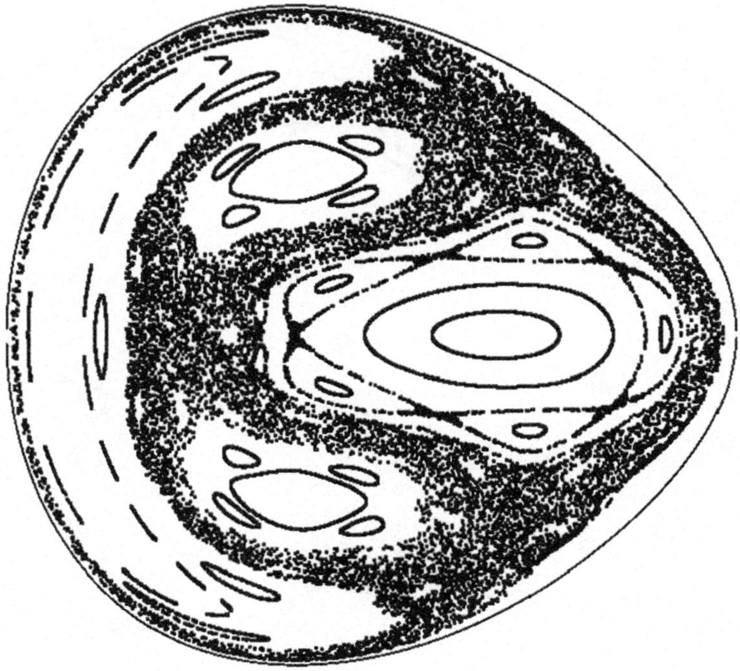

Fig. 1.21 Surface of Section of the Hénon-Heiles system for energy 0.12 showing the formation of chains of libration islands

them, new asymptotic branches originate and give rise to libration zones or chaotic entanglements, depending on whether the perturbed system remains integrable or ceases to be so.

In Fig. 1.21, we show one Surface of Section of the Hénon-Heiles system for an energy value different from those of Fig. 1.10 (0.12) and chosen for the rich features it shows. The chain of 5 islands seen in the central part of the figure is an illustration of the processes described in this section. They enclose a periodic orbit that crosses the section 5 times before closing and are surrounded by a stochastic layer created by the entanglement of the asymptotic branches emanating from the 5 corresponding saddles. Similar phenomena are also seen in other parts of the section.

The classical version of Poincaré's theory is completed by recalling that the set of periodic solutions of an integrable system is dense. At each point on the Surface of Section, we are infinitely close to periodic orbits, that is, to invariant tori on which solutions close after a certain number of iterations. To each of such tori, we can associate a rational number formed by two integers prime to each other. One of these integers is the number q, which gives the number of turns made by the periodic solution, longitudinally to the torus, before closing. But at the same time, the torus is also being circled transversely. And as the solution closes after one period, the torus should have been transversely circled an integer number p of times. We can then associate the rational number p/q to the periodic solutions on the considered torus. It is obvious that p and q are prime to each other; otherwise, the periodic

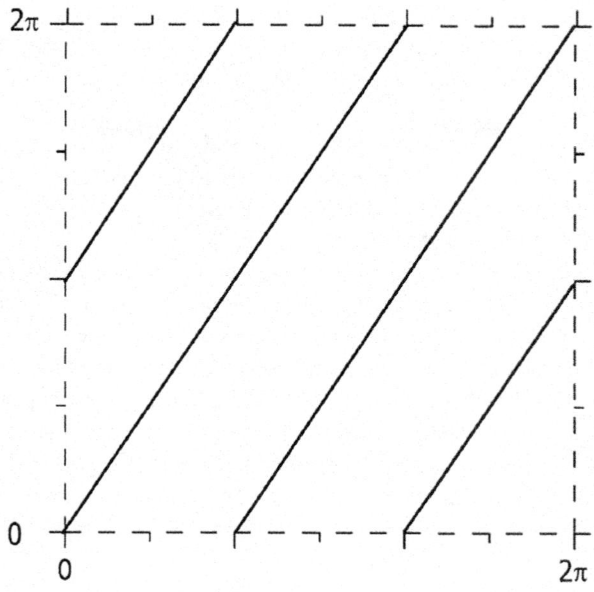

Fig. 1.22 Planar representation of a periodic orbit that circles the torus twice in one direction and three times in another before closing

solution would have closed much earlier. To illustrate this fact, we should draw a torus and the periodic orbit around it. This is not an easy task, and the attempts we see in many books are generally unsatisfactory. We will therefore do it using an artifice. First, we section the torus transversely, transforming it into a cylinder; then, we cut this cylinder along a height, transforming it into a rectangle (or square). This operation allows us to see, in a simple way, the entire surface of the torus (Fig. 1.22).

This result has an interesting and important consequence for what follows. If the periodic orbits are associated with rational numbers, the nonperiodic ones are associated with irrational numbers. In the latter case, there are not two numbers p and q such that the orbit closes after these numbers of transverse and longitudinal circlings of the torus. But they always are as close to the periodic ones as one wants.

1.20 The KAM Theory (Kolmogorov - Arnold - Moser)

The analyses done by Poincaré and Birkhoff showed that when an unstable system is perturbed, stochastic layers can appear throughout the phase space. However, it is not stated that they must appear. For instance, if the resulting system is integrable, no stochasticity will appear[8]. The best knowledge about the effects of adding a

[8] A funny way of expressing the exceptionality of these systems is to say that integrable systems only exist in books!

perturbation to an integrable system is given by the KAM theory, whose main result was announced by Andrei Kolmogorov in 1954 and whose basic theorems were proved in detail by Vladimir Arnold and Jürgen Moser in papers published between 1963 and 1966.

We will illustrate it by having in mind the Surface of Section of a system with two degrees of freedom subjected to a perturbation. More particularly, we will consider the simple case of a perturbation added to an integrable system with a stable periodic orbit at the center surrounded by invariant tori.

In general, the integrability is not preserved by the perturbation, and the stochastic layers predicted by the Poincaré-Birkhoff theory appear. The corresponding invariant tori are destroyed by the perturbation added to the system. In its place appear sets of islands around centers and saddle points from which asymptotically stable and unstable branches emanate and later intersect, forming chaotic tangles. There will be regions of space in which bundles of invariant tori of the integrable system are destroyed, giving rise to a stochastic layer (Fig. 1.23). Motions in this region are extremely complicated and clearly irregular. This can occur in several places. But for some invariant tori, it may not occur. The KAM theory says the

Fig. 1.23 Stochastic layers separated by KAM tori (Reprinted from Arnold et al. (1988). Copyright 1988 Springer)

following: If the integrability of the system is not preserved by the perturbation, anywhere in the phase space, we will be infinitely close to a stochastic layer in which the invariant tori were destroyed. However, if the perturbation is small enough, the volume of the phase space occupied by the preserved invariant tori (the so-called KAM tori) is non-zero. Increasing the perturbation, the volume occupied by the preserved tori decreases[9].

To visualize this measure, in the case of a Surface of Section, one can consider a segment transverse to the KAM tori sections. If we divide this segment into N minute intervals and call n the number of intervals intercepted by at least one KAM torus, the KAM theorem tells us that when the size of the intervals tends to zero, the fraction n/N tends to a non-zero value. This means that, no matter how fine the division is, there is always a huge number of intervals intersecting an infinity of KAM tori.

The KAM theory, however, does not say how much of the tori of the original phase space is destroyed and how much is preserved! Indeed, if the perturbed system is not integrable, we will always be close to stochastic layers resulting from the destruction of the invariant tori of the original system. When we analyze a system with two degrees of freedom through Surfaces of Section, we get regions where destruction is evident, but there are also regions where the motion seems regular. It is a problem of graphic discernment. A powerful magnification should show that this regular-looking region contains numerous stochastic layers between the invariant surfaces, and they would be seen in as large a number as large is the magnification used.

The Surface of Section in the region where there are preserved invariant surfaces looks like the cut of an onion. The layers of the onion are the stochastic layers, and the interstitial skins separating one layer from another are the invariant surfaces preserved by the perturbation. But it has to be a hypothetical onion of infinitely many layers because the stochastic layers form from periodic solutions of the integrable system, and periodic solutions form a dense set within the set of orbits corresponding to a given energy. This kind of structure has an important consequence in systems with two degrees of freedom. The preserved invariant tori confine the chaotic motion occurring in stochastic layers. Solutions that enter or leave the stochastic layer cannot exist because the invariant tori cannot be crossed (a solution that has a point on an invariant surface must remain on it forever). In this case, the chaos exists but is confined.

If the perturbed system is no longer integrable, there is no other law of conservation in this system besides the law of conservation of energy. But there is a quantity that will be conserved by a solution when it is on a KAM torus. In addition, between two KAM tori, this quantity can only vary between the values corresponding to these tori. Thus, even if this quantity is not conserved along a chaotic solution, if the KAM tori are very close to one another (such that the cut by

[9] This volume has to be given by a Lebesgue integral because KAM tori do not touch each other. Between any two preserved invariant tori, there is at least one stochastic layer.

a Surface of Section looks like the cut through an "onion" of extremely thin layers), the confinement of the solution limit its value, so that it may not be possible to verify any variation experimentally. A two-degree-of-freedom system with microscopic stochastic layers behaves, from a physical point of view, as if it were an integrable system.

References

Arnold, V.I., Kozlov, V.V., Neishtadt, A.I.: Dynamical Systems III. Springer, Berlin (1988)
Celletti, A.: Stability and chaos in Celestial Mechanics. Springer, Berlin (2010)
Contopoulos, G.: Order and Chaos in Dynamical Astronomy. Springer, Berlin (2002)
Contopoulos, G., Polymilis, C.: Geometrical and dynamical properties of homoclinic tangles in a simple Hamiltonian system. Phys. Rev. E **47**, 1546–1557 (1993)
Ferraz-Mello, S.: Canonical Perturbations Theories. Degenerate Systems and Resonance. Springer, New York (2007)
Ferraz-Mello, S., Beaugé, C., Folonier, H., Gomes, G.O.: Tidal friction in satellites and planets. The new version of the creep tide theory. Eur. J. Phys. **229**, 1441–1462 (2020)
Ferraz-Mello, S., Dvorak, R.: Chaos and secular variations of planar orbits in 2: 1 resonance with Dione. Astron. Astrophys. **179**, 304–310 (1987)
Ford, J., Stoddard, S. D., Turner, J. S.: On the integrability of the Toda lattice. Progr. Theor. Phys. **50**, 1547–1560 (1973)
Guckenheimer, J., Holmes, P.: Nonlinear Oscillations, Dynamical Systems, and Bifurcations of Vector Fields. Springer, New York (1983)
Hénon, M., Heiles, C.: The applicability of the third integral of motion: some numerical experiments. Astron. J. **69**, 73–78 (1964)
Jeans, J.: Astronomy and Cosmogony. Dover, New York (1961)
Kurth, R.: Introduction to Stellar Statistics. Pergamon, Oxford (1967)
Laskar, J.: A numerical experiment on the chaotic behaviour of the solar system. Nature **338**, 237–238 (1989)
Laskar, J.: Les solutions astronomiques pour l'étude des paléoclimats (2017). www.iap.fr/vie_scientifique/ateliers/Astronomie_et_Climat/2017/videos/Jacques_LASKAR_2017-06-14_1430/index.html
Lichtenberg, A.J., Lieberman, M.A.: Regular and Stochastic Motion. Springer, New York (1983)
Poincaré, H.: Les Méthodes Nouvelles de la Mécanique Céleste, vol. 3. Gauthier-Villars, Paris (1899)
Sussman, G.J., Wisdom, J.: Chaotic evolution of the solar system. Science **257**, 56–62 (1992)
Wisdom, J.: Urey Prize Lecture: chaotic dynamics in the solar system. Icarus **72**, 241–275 (1987)
Yokoyama, T.: Reappearance of ordered motions in strongly nonlinear Hamiltonian systems. Celest. Mech. **33**, 99–109 (1984)

Chapter 2
Resonant Asteroidal Dynamics

2.1 Resonant Asteroids

Asteroids are small bodies, most of them orbiting in the space between the orbits of Mars and Jupiter (Fig. 2.1). The largest of them all, Ceres, was discovered by Giuseppe Piazzi, in Palermo, Sicily, on January 1st, 1801, the first day of the nineteenth century, with a diameter of 930 km (for reference, remember that the Moon has a diameter of 3480 km). Ceres was the first asteroid discovered and is currently classified as a dwarf planet. After Ceres, many others were discovered, but they were much smaller. Around 1990, 20 thousand were known, most of them with diameters in the range of 1–30 km, with only a few of them with diameters greater than 200 km.

In 1980, the geologist Walter Alvarez and his father, the physicist Luiz Alvarez, studying samples of a sediment layer in Gubbio, Italy, found an abnormal concentration of Iridium, which they assumed to be of extra-terrestrial origin (Alvarez 2008). The sediment layer sampled was exactly at the K-T boundary, the thin layer that separates the sediments of two geological eras, the Cretaceous and Tertiary. When similar discoveries were made in other outcrops of the K-T boundary, in various parts of the world, it became evident that this material must have originated from a global catastrophe resulting from the collision with the Earth of a large asteroid, with a diameter between 10 and 15 km, at the time that layer was formed, 65 millions of years ago.[1] This discovery and the observation of the existence of many other craters resulting from large meteorite strikes on the Earth's surface have shown that the possibility of a major catastrophe caused by an asteroid colliding with the Earth is

[1] The crash site was later identified as a crater covered by sediments in the Chicxulub region of the Yucatán Peninsula, Mexico. The global catastrophe that created this crater has been associated with the extinction of the dinosaurs and some 75% of the plant and animal species then existing on Earth, and with the subsequent development of mammals, among them the primitive primates, ancestors of humans.

S. Ferraz-Mello, *Chaotic Dynamics in Planetary Systems*,
Astronomy and Planetary Sciences, https://doi.org/10.1007/978-3-031-45816-3_2

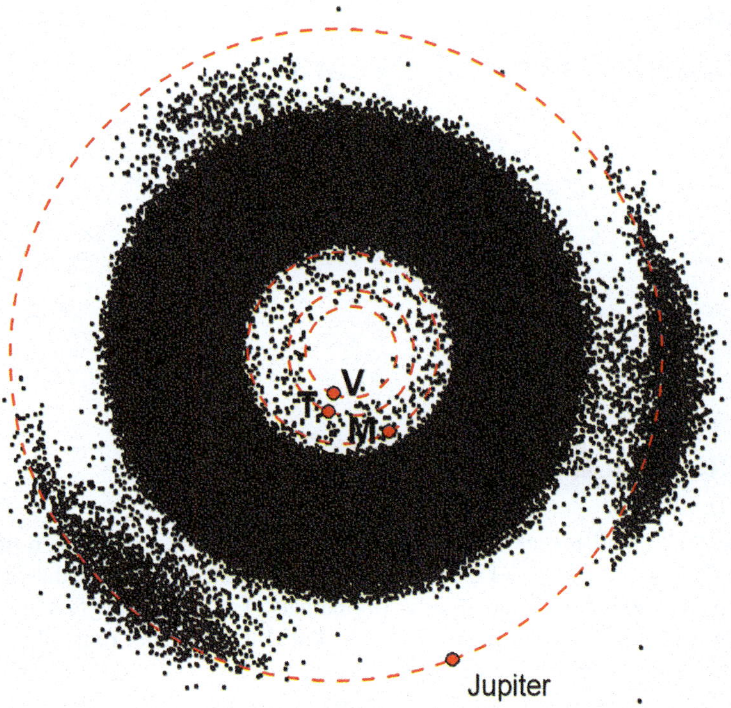

Fig. 2.1 Positions of 526,000 asteroids and the planets Jupiter, Mars (M), Earth (T) and Venus (V) on May 31st, 2020. Shown are the asteroids in the main belt, those orbiting close to Jupiter's orbit (called Trojan asteroids), and the asteroids approaching the Earth (the Near-Earth asteroids or NEAs). The dashed lines show the orbits of the 4 planets included in the figure. The perihelion of Jupiter's orbit is at longitude 13.9 degrees, on the right side of the figure. Source: Lowell Observatory astorb database

real. Although of low probability, the assessment of the potential damage has led to assigning a high priority to the discovery of all asteroids that approach the Earth and are capable of causing major damage in the event of a collision. Several telescopes have been built for this purpose, and the number of known asteroids has increased rapidly. It reached 60 thousand at the turn of the century and is now approaching one million, of which more than 20 thousand have orbits that approach the Earth.

The analysis of the distribution of asteroid orbits shows many important characteristics and reveals much of the past evolution of this collection of celestial bodies. The most spectacular of these, discovered in 1866 by the astronomer and mathematician Daniel Kirkwood, is the existence of gaps in the distribution of their orbital periods, in the exact positions that would correspond to asteroids with periods commensurable with that of Jupiter with ratios given by simple fractions. Figure 2.2 shows the distribution of the orbits of the asteroids according to their average distance from the Sun, representing in the ordinates the number of asteroids per 100,000 km interval.

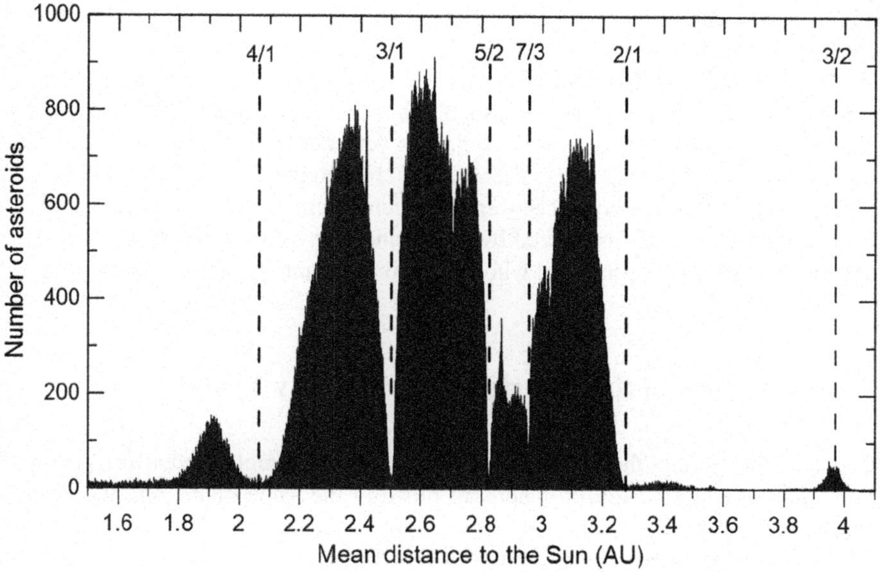

Fig. 2.2 Distribution of the orbits of the asteroids in the main belt according to their average distance from the Sun. The dashed lines indicate distances from the Sun that correspond to orbits with periods commensurable with that of Jupiter at the indicated ratios

At the time of Kirkwood's study, only 91 asteroids had been discovered. The visible gaps in Fig. 2.2 became evident to astronomers after the discovery of the asteroids (65) Cybele and (76) Freia, the first to be found at a distance from the Sun greater than 3.42 AU,[2] while the range between 3.15 and 3.4 AU was completely unpopulated. The discovery of another asteroid in this situation: (87) Sylvia, in May 1866, must have motivated Kirkwood to study the phenomenon. He was the first to try to justify these gaps by the fact that they correspond to orbital periods commensurable with the orbital period of Jupiter and to compare them to the Cassini divisions observed in Saturn's rings. These gaps in the distribution of asteroids are now known as "Kirkwood gaps".

A contrasting phenomenon to the Kirkwood gaps was discovered some years later. In 1875, the Austrian astronomer Johann Palisa discovered the asteroid (153) Hilda in an orbit whose period was almost exactly 2/3 of the period of Jupiter. In 1888, the same Palisa discovered (279) Thule, this one in an orbit whose period was almost exactly 3/4 of the period of Jupiter. Time has shown that at least the first of these asteroids has nothing exceptional: today, more than 4000 asteroids are known whose periods are very close to 2/3 of the period of Jupiter. And this number tends to increase as observations allow the discovery of less bright objects. Thule was, for

[2] The astronomical unit AU is a unit of length very close to the mean distance from the Earth to the Sun. 1 AU = 149,597,870 km.

a long time, the only one in its class. Today two other small asteroids with the same characteristics are known.

These groups are of considerable importance, for any theory which proposes to explain the Kirkwood gaps must be such that it does not also empty the resonances where a group of asteroids is observed. It is very common to read here and there that the Kirkwood gaps are due to resonances. Indeed they are, and this will be seen in this chapter. But, if resonances were sufficient to create the gaps, there should be no asteroids like those discovered by Palisa and many others afterward, which are almost certainly primordial, and whose orbit is resonant.

2.2 Asteroids in the Restricted Three-Body Model

Let us begin to study the motion of asteroids with the simple model formed by the Sun, Jupiter, and an asteroid, where all move in the same plane, and the orbit of Jupiter around the Sun is circular. This problem is called the "Restricted (circular) 3-body problem". How many degrees of freedom does this system have? To solve this question, we take a system of axes uniformly rotating with the same angular velocity as that of the orbital motion of Jupiter. Since, in this problem, Jupiter's motion is circular, the position of Jupiter is a fixed point in this referential (we usually choose the axes in such a way that Jupiter is on the x-axis as in Fig. 2.3). Then, the configuration of the system at any given moment is fully known if the two coordinates of the asteroid in this referential are known. Therefore, the system has two degrees of freedom and can be studied with the help of Surfaces of Section as discussed in Sect. 1.12.

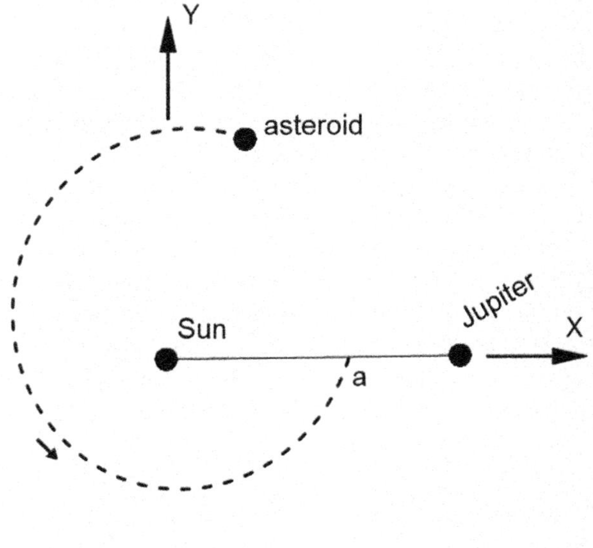

Fig. 2.3 Position of the asteroid in a coordinate system whose axes rotate around the Sun with the same angular velocity as Jupiter. The asteroid moves along its orbit and intersects the Sun-Jupiter line at point a

To construct the Surfaces of Section, we simulate numerically the motion of the asteroid from a chosen initial position until its path crosses the Sun-Jupiter line (point a in Fig. 2.3). The coordinates of the first point of the Surface of Section are the distance of the asteroid from the Sun and its radial velocity (i.e. $x_a, \dot{x}_a$). Next, we continue the simulation until the trajectory again crosses the Sun- Jupiter line at a new point b (which do not coincide with a because of the perturbations of the motion of the asteroid due to the gravitational attraction of Jupiter), and so on. Repeating the process for different initial conditions, all of them corresponding to the same energy level, and superposing the results, we have a Surface of Section (or Poincaré map) of the studied system.

In the case of the asteroidal model of 3 bodies, centers appear which correspond exactly to the resonances between the motion of the asteroid and Jupiter. Let us consider, as an example, the case of asteroids whose initial state lies within the 2:1 resonance with Jupiter. To allow a better interpretation of the results, we will transform them into orbital parameters of known significance. These are the eccentricity and the critical angle $\theta = \lambda_{\text{ast}} - 2\lambda_{\text{Jup}} + \varpi_{\text{ast}}$ (the λ are mean longitudes and ϖ_{ast} is the longitude of the perihelion of the asteroid). The results are shown in Fig. 2.4 in two cases corresponding to different eccentricities. In both cases, we see a family of oscillatory motions of the angle θ around $\theta = 0$ (center of the resonance). Similar figures can be obtained using the semi-major axis as a parameter instead of the eccentricity. Within asteroidal resonances, the eccentricity and the semi-major axis have strongly correlated variations, and one of them increases when the other decreases.

The centers are stable periodic orbits. Let us give a simple kinematical description of these periodic orbits. First, however, let us introduce a simplification of language. The orbital period of Jupiter is 11.86 years (11 years, 10 months, and

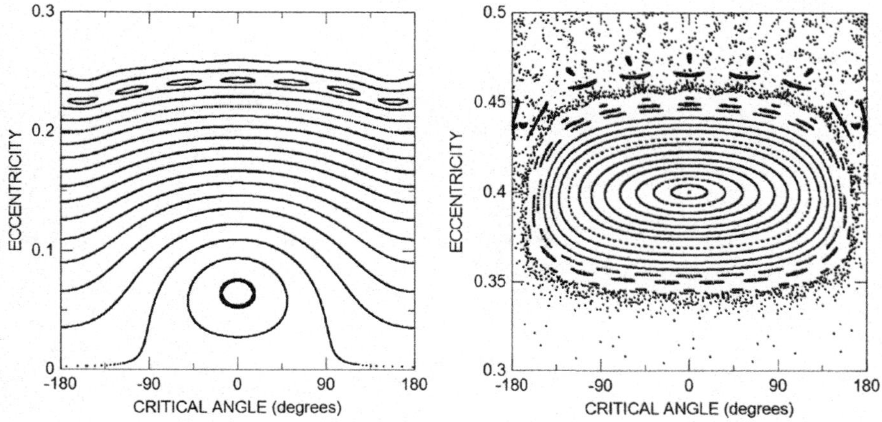

Fig. 2.4 Surfaces of Section of the motion of an asteroid centered on the resonance 2:1 with Jupiter, at low (left) and high (right) initial eccentricities, in the restricted model with Jupiter in a circular orbit

10 days). To simplify, we will round Jupiter's orbital period to 12 years. 12 is a convenient number since it is divisible by 2, 3, and 4. In this simplification, if an asteroid has an orbital period of 9 years, 8 years, 6 years, 4 years, or even 3 years, it will be moving in resonance with Jupiter.

2.2.1 *The Resonance 2:1*

Let us begin with the case of an asteroid whose orbital period is half the orbital period of Jupiter, that is, with the simplification introduced above, a period of 6 years. As the asteroid is in a more internal orbit than Jupiter's, its motion is faster. There are times when the asteroid is in conjunction with Jupiter (that is, in a Sun-Asteroid-Jupiter alignment). Let us study the motion starting from the conjunction (Jupiter and the asteroid in the positions 1 in Fig. 2.5). After 6 years, the asteroid returns to its initial position. When the asteroid completes two turns, Jupiter will have completed one turn. Marking the positions of the asteroid and Jupiter every 3 years, we have the positions 1, 2, 3, and 4, then return to 1. In 12 years they both return to the same initial position, aligned with the Sun (and on the same side). We note that there are no other occasions in this period with the asteroid in conjunction with Jupiter (that is, in the Sun-Asteroid-Jupiter alignment). They only meet, one in front of the other, at the positions 1 of their orbits.

Let us now consider the real mass of Jupiter, which is about 1/1000 of the mass of the Sun. Every time the asteroid passes in front of Jupiter, it is attracted by the

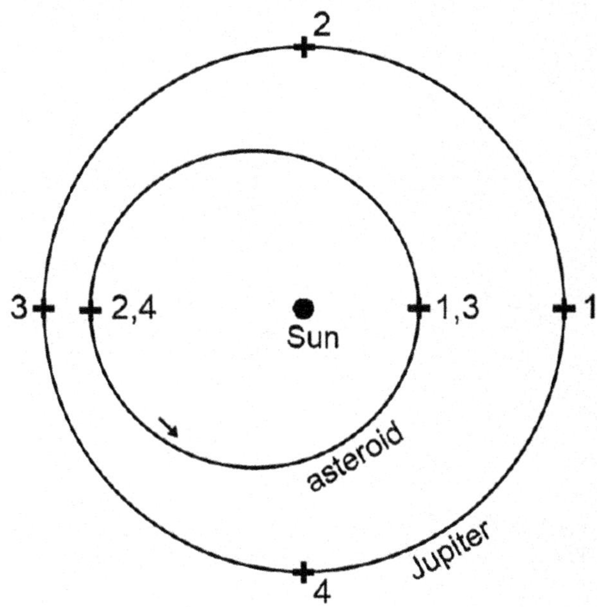

Fig. 2.5 Periodic orbit of the asteroid in the resonance 2:1 with Jupiter in circular orbit and that of the asteroid with eccentricity $e = 0.3$

planet with greater intensity and has its energy, and therefore its orbital period, changed (see Sect. 2.3). But, if the asteroid is in an orbit like the one described above, the symmetry of the asteroidal motion with respect to the line where the conjunction with Jupiter occurs implies that the final balance of energy exchanges is null and the asteroid can remain in the same orbit. In Fig. 2.5, in the initial symmetric conjunction, the asteroid is in the perihelion and the orbit is stable. In this periodic stable orbit, at the moment of the conjunction, the asteroid is as far away from Jupiter as possible. Reversing the situation and placing the asteroid in the aphelion at the time of the initial conjunction, the orbit is unstable. In the periodic stable orbit, at the moment of the conjunction, the asteroid is far from Jupiter, so the perturbing effect of Jupiter's attraction is minimal. In the unstable periodic orbit, the opposite happens.

On the Surfaces of Section, the stable orbit corresponds to the centers seen in the middle of Fig. 2.4. The closed curves, which appear around these centers, indicate that, to remain in the resonance, the asteroid does not need to be exactly in the perihelion at the time of the conjunction but has to be close to it. To understand this, let us suppose that the conjunction occurs a little before the asteroid reaches the perihelion of its orbit; in this case, we no longer have the symmetry of motion that guarantee the periodicity of the orbit; the action of Jupiter, although minimized by the great distance, will change the energy of the asteroid (as in a swing-by maneuver; see Sect. 2.3) and, as a consequence, its orbital period will suffer a slight increase; therefore, the asteroid will decrease its velocity and, the next time, the conjunction will occur a little closer to the passage of the asteroid through the perihelion. And so on, until the asteroid is in the perihelion at the time of the conjunction. However, as it had its orbital period continuously increased, it will be moving with a period longer than that corresponding to the exact commensurability of the periods. Therefore, in the following turns, the asteroid will continue moving more slowly, and the conjunction will occur after the asteroid passes by the perihelion. The situation, at the moment of the conjunction, is reversed and now the action of Jupiter will tend to decrease the orbital period of the asteroid. When the period reaches the value corresponding to the commensurability, the position of the conjunction, which was moving away from the perihelion, will move toward it. Therefore the resulting asteroidal motion is such that the position at which the conjunction of the asteroid and Jupiter occurs oscillates periodically around the perihelion of the asteroid's orbit. This motion is called "libration". During a period of the libration, there are conjunctions in which the asteroid is before the perihelion, and there are conjunctions in which the asteroid is after the perihelion. Like a pendulum whose stable equilibrium point corresponds to a conjunction in which the asteroid is exactly in the perihelion of its orbit.

At high eccentricities (Fig. 2.4 right), if the amplitude of the variation of the conjunction position is too large, the motion of the asteroid reaches the chaotic entanglement associated with the unstable periodic orbit and stops librating. The position of the conjunction with Jupiter will now be distributed over the whole orbit. The asteroid is no longer in resonant motion.

2.2.2 The Resonance 3:2

Figures like the one for resonance 2:1 can be constructed for other resonances. Let us consider, for instance, the case of the asteroidal resonance 3:2, that is, the asteroids with periods close to 8 years. In this case, the average angular velocity of the asteroid is 1.5 times that of Jupiter. So if we observe the two motions, the asteroid starts faster, but its advance over Jupiter progresses slowly, and when Jupiter completes 2/3 of one turn, the asteroid returns to its initial position. As in the figure of the previous case, let us mark the sequences of points, but every 4 years: 1, 2, 3, 4, 5, 6 and return to 1. If the asteroid's orbit is elliptical, the attraction of Jupiter at the conjunction (Sun-asteroid-Jupiter alignment) is minimized when the conjunction occurs in the asteroidal perihelion (point 1 on the asteroid's orbit) and maximized when the conjunction occurs in the asteroidal aphelion (point 2 on the asteroid's orbit).

Some congruencies appear in Fig. 2.6, which should be noted: In a stable orbit, when the asteroid is in its orbit's perihelion, Jupiter can be in positions 1, 3, or 5. Hence, in this case, the solution to the problem is not unique. If the orbit of the asteroid is rotated of 120°, we have the same relative situation as before the rotation. In these cases, when the asteroid moves farther away from the Sun (aphelion), and thus comes dangerously close to the orbit of Jupiter, the planet is in one of the positions 2, 4, and 6 indicated in the figure. That is, in the stable periodic orbits, the asteroidal aphelion is either on the opposite side of Jupiter or 60° distant from the planet. The existence of a conjunction close to the perihelion continues to determine the stability of the asteroid's libration.

To show how the asteroids in stable periodic orbits would be distributed in space, we present, on the right-hand side of Fig. 2.6, the stable periodic orbit in a moving reference system. Instead of a system with fixed axes, we have a system whose

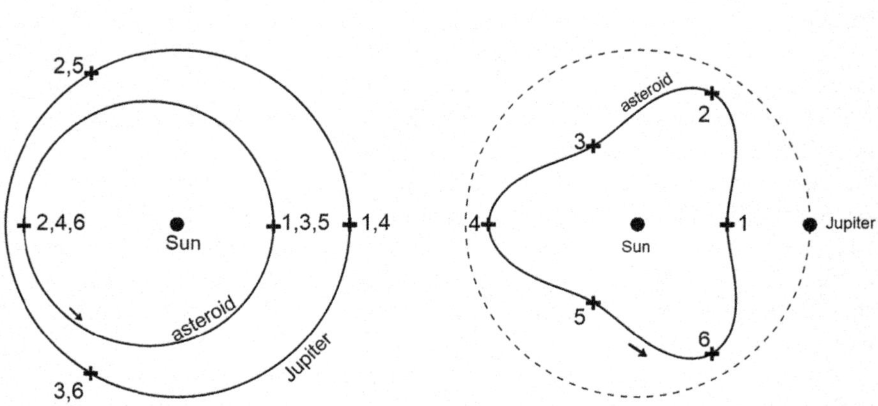

Fig. 2.6 Periodic orbit of the asteroid in resonance 3:2 with Jupiter in circular orbit and one asteroid with eccentricity $e = 0.25$. Left: Representation in a system of axes with fixed directions in space; Right: Representation in a system of axes rotating with the same angular velocity as Jupiter (rotating system in which Jupiter stands still)

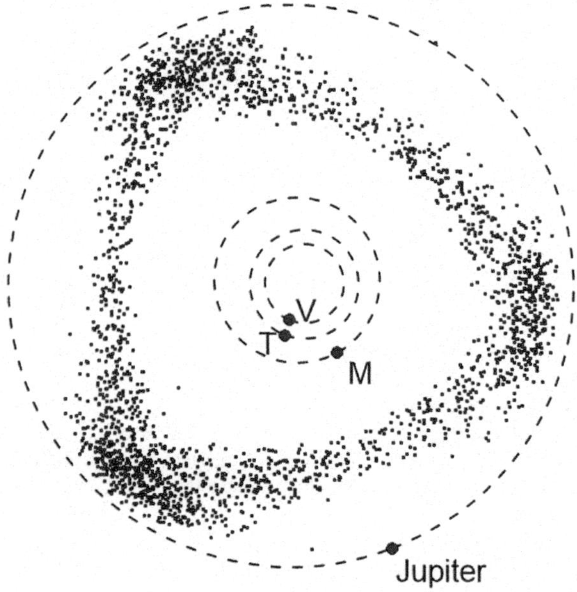

Fig. 2.7 Position of the asteroids of the 3:2 resonance on May 31st, 2020. The dashed lines show the orbits of Venus, Earth, Mars, and Jupiter. Source: Lowell Observatory astorb database

axes are revolving around the Sun with the same angular velocity as Jupiter so that Jupiter remains stationary in this system. The asteroid describes a curve of 3 lobes, as shown in the figure, and Jupiter is represented by one point.

The asteroids moving in the vicinity of the periodic stable orbit are seen in Fig. 2.1, which shows the distribution of all asteroids with well-determined orbits. In it, we mainly see the asteroids of the main belt, but we also see the 7000 asteroids known as Trojans, which are also resonant and have the same orbital period as Jupiter (nearly 12 years) and which are always at a distance from Jupiter that oscillates around 60° (in fact, oscillating asymmetrically so that Trojan asteroids may be found at distances between 40° and 90° from Jupiter). The asteroids of the 3:2 resonance moving close to the aphelions of their orbits are also noticeable because they are moving far from the main belt. They may be found in Fig. 2.1 between the main belt and the Trojan asteroids, at some 60° distant from Jupiter (on both sides), and between the main belt and Jupiter's orbit on the side opposite to Jupiter (i.e., at 180° from Jupiter). The others are moving inside the main belt and cannot be identified in the figure. To see them, we can redraw the figure considering only the 2380 asteroids whose orbits are in the resonance 3:2 (Fig. 2.7)

2.2.3 The Resonance 3:1

Let us now consider the case of the asteroidal resonance 3:1, that is, the asteroids with orbital periods close to 4 years. In this case, the average angular velocity of the

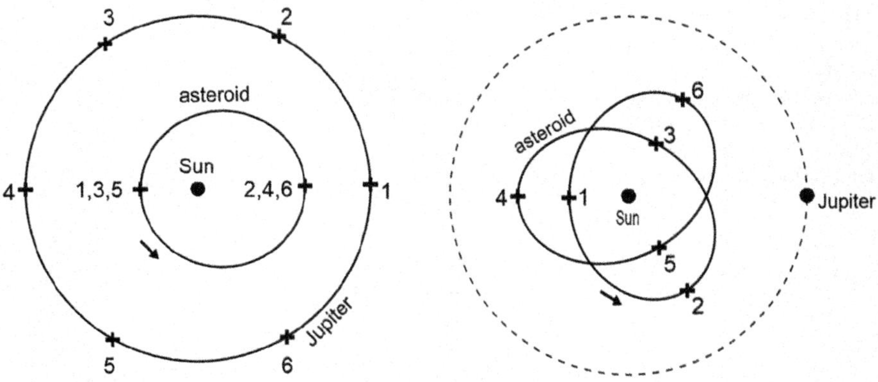

Fig. 2.8 Periodic orbit of the asteroid in resonance 3:1 with Jupiter in a circular orbit and the asteroid with eccentricity $e = 0.3$. Left: Representation on a system of axes with fixed directions in space; Right: Representation on a system of axes rotating with the same angular velocity as Jupiter (rotating system in which Jupiter stands still)

asteroid is three times that of Jupiter. So, if we observe the two motions, the asteroid goes faster, and its advance over Jupiter progresses rapidly. When Jupiter completes 1/3 of turn, the asteroid is already back to its initial position. As in the figure of the previous case, let us mark the sequences of positions occupied by the asteroid and by Jupiter, every 2 years: 1, 2, 3, 4, 5, 6, and return to 1.

Let us note that in the orbit shown on the right-hand side of Fig. 2.8 showing the motion of the asteroid in the rotating system, the asteroid makes two turns before returning to the starting position. Another point to note is that, in the initial position 1, the asteroid and Jupiter are aligned with the Sun, but not on the same side. They are in opposition with respect to the Sun. As this resonance is of second order,[3] a conjunction in the perihelion would lead, 1.5 turns later, to a conjunction in the aphelion in which the asteroid and Jupiter would be aligned with the Sun and on the same side, a proximity that, in general, does not allow a periodic orbit to be stable.

Finally, note that, in this case, if we start counting the angles, for example, in the points 1, with the asteroid in the perihelion of its orbit, we have $\lambda_{\rm Jup} = 0°$, $\lambda_{\rm ast} = 180°$ and $\varpi_{\rm ast} = 180°$. Therefore, the critical angle $\theta = \lambda_{\rm ast} - 3\lambda_{\rm Jup} + 2\varpi_{\rm Jup}$ is equal to 540° which, in the trigonometrical circle, is equivalent to 180°. In the periodic orbit shown, the critical angle remains constant throughout the motion.

As a common conclusion to the three cases discussed, we can say that in the restricted circular model, the resonant asteroids can be, in general, well protected. If the Solar System were formed simply by the Sun, one planet like Jupiter in a circular orbit, and the asteroids, all the above-discussed resonances should be full of asteroids. In all cases, stable periodic orbits exist and asteroids in these orbits, or in

[3] The order of a resonance is the difference between the two integers that form the rational number that represents it. The order of a resonance $p : q$ is equal to $p - q$. The order of the resonance 3:1 is $3 - 1 = 2$.

neighboring orbits, could keep moving forever in this condition, always protected from a more violent gravitational interaction with Jupiter.

2.2.4 *The Pluto-Neptune Resonance*

A problem of the same nature as the ones above is that of the motions of Pluto and some objects in the Kuiper belt. However, in this case, the small bodies—Pluto or the Kuiper belt objects—are in orbits external to the orbit of the large planet disturbing their motions. The period of Neptune's motion around the Sun is 165 years, and that of Pluto is 248 years, that is, 3/2 of Neptune's period. The average angular velocity of Pluto is 2/3 of that of Neptune (unlike the asteroids, Pluto moves more slowly than Neptune). Thus, if we observe the two motions, Neptune starts moving ahead and its advance over Pluto progresses slowly, and when Neptune returns to its initial position, Pluto completed 2/3 of a revolution. As in the previous figures, let us mark in Fig. 2.9 the sequences of points every 83 years: 1, 2, 3, 4, 5, 6, and return to 1. As Pluto's orbit is elliptical, the attraction of Neptune at the conjunction is minimized when it occurs when Pluto is in the aphelion (point 1 of its orbit) and maximized when the conjunction occurs in Pluto's perihelion. (There is an inversion with respect to the case of the asteroids, which are internal to Jupiter, and which protect themselves from Jupiter when they avoid conjunctions near the asteroidal aphelion). It is easy to see that when Pluto is in the perihelion of its orbit, halfway between 2 and 3 (or 5 and 6), Neptune will be 90° ahead or behind Pluto. The indicated positions of Neptune and Pluto in the figures are those of May 2020.

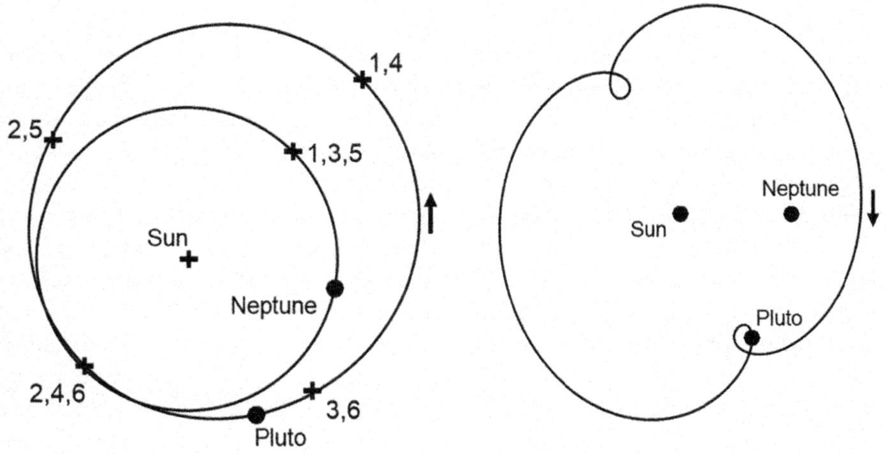

Fig. 2.9 Orbits of Neptune and Pluto. Left: Orbits in a system of axes with fixed directions in space; Right: Representation on a system of axes rotating with the same angular velocity as Neptune (in this rotating system Neptune is almost immobile)

To better see the relative motion of Pluto and Neptune, it is convenient to redraw the orbit of Pluto by changing the reference system. Instead of a system with fixed axes, we take a system of axes rotating around the Sun with the same angular velocity as Neptune so that Neptune appears stationary in this reference system. In the case of several other objects in 2:3 resonance with Neptune (the so-called Plutinos), the orbit in the rotating frame looks like an elongated ellipse squeezed at its central part. In the case of Pluto (and of some other Plutinos), it is a little more complicated because the orbital eccentricity is very large (0.248), and the orbit crosses that of Neptune. If, in general, Pluto moves slower than Neptune, it will move faster than Neptune when it is in the interior of Neptune's orbit. Therefore, in the system of axes rotating with Neptune, Pluto reverses the direction of its motion several times and thus draws a double-looped curve (Fig. 2.9 right). In this figure, we see that when Pluto enters Neptune's orbit, it is safe from a greater disturbance by Neptune because it will be far away from the planet, ahead or behind it. As the difference in velocity is small near the perihelion of its orbit, the motion of Pluto in the loops of its relative orbit is very slow. Compare the large arc traveled by Pluto, since it passed through the perihelion of its orbit (Fig. 2.9 left), to the small arc within the loop of its orbit in the figure on the right. The motion of Pluto has a libration of the critical angle $\theta = 3\lambda_P - 2\lambda_N - \varpi_P$ around 180° and Pluto is currently some 20° ahead of the corresponding position in the stable periodic orbit. As a consequence, in the right-hand graph, the orbit appears skewed. In the stable periodic case, the orbit is symmetric about the Sun-Neptune axis, and the critical angle is $\theta = 180°$, exactly (see Gallardo and Ferraz-Mello 1998).

2.3 Close Encounters. Swing-by

Let us consider an asteroid in motion around the Sun. During its motion, it may approach one of the planets. The effect of one close encounter is a change of the direction and energy of the asteroid's motion. The dynamics is the same as in the gravity-assisted maneuver known in Astronautics as swing-by (or slingshot) used to change the course and speed of a space probe targeting a distant planet. It allows a space probe to leave the Earth with less energy than is necessary to accomplish its interplanetary mission; on its way, it has a close encounter with a planet (which can be the Earth itself) and draws from this encounter the mechanical energy necessary to reach its target. If the probe has an excess of energy, this excess will be used to travel the rest of the way with greater speed and arrive more quickly. Additionally, if the mission requires the probe to go into orbit around the planet, it will be necessary, upon arrival, to subtract the excess energy; again, this can be done by means of a close encounter with a large satellite of the planet visited, so that the probe has decreased its mechanical energy.

Consider the Sun, one planet in a circular orbit around the Sun, and an asteroid (or space probe) in the plane of planet's orbit, which has a close encounter with the planet. This mechanical system has 2 degrees of freedom (only the coordinates of the asteroid are unknown) and a peculiar conservation law, called the Jacobi integral. We will write it in heliocentric form

$$\mathcal{J} = E - nS + W_p = \text{const.} \tag{2.1}$$

where E is the heliocentric orbital energy of the asteroid (the sum of its kinetic energy and the potential energy of its gravitational attraction by the Sun), n is the angular velocity of the planet in its motion around the Sun, S is the angular momentum of the asteroid, considered positive when the asteroid and the planet are moving around the Sun in the same direction, and negative when they move in opposite directions, and W_p is the potential energy of the gravitational attraction of the asteroid by the planet (Brouwer and Clemence 1961, p. 256). The constant $\mathcal{J}$ is the so-called Jacobi constant.

We will explain in a very simplified way what happens when an asteroid coming in an elliptical orbit passes close to one planet. Let us draw a circle around the planet and, let us suppose that the attraction of the asteroid by the planet is negligible outside the circle (which represents the sphere of influence of the planet). Let us mark several points on this circle: Points A and B are the points at which the asteroid enters and leaves the circle. Point B' is the point through which the asteroid would leave the circle if it were not deviated from its initial orbit by the planetary attraction. Figure 2.10 describes close encounters in which the speed of the asteroid relative to the planet is much greater than the escape velocity of the planet. In Astronomy,

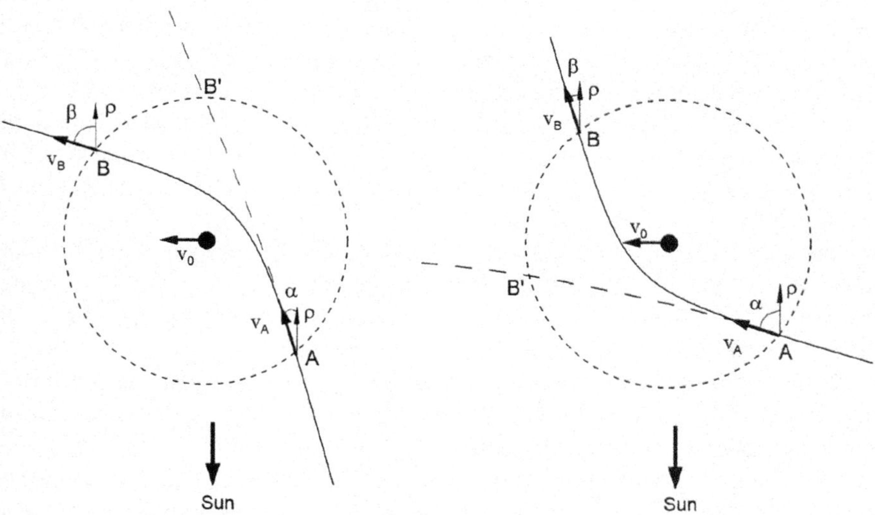

Fig. 2.10 Trajectory of an asteroid (or comet) passing at high speed through the sphere of influence of a planet. Left: Passing behind the planet. Right: Passing in front of the planet. Dashed line: The path that it would follow if there were no encounter. Vectors: **v**: Unit vectors of the velocities of the asteroid at A and B and of the velocity of the planet; $\boldsymbol{\rho}$: unit vector of the direction opposite to the direction of the Sun

these encounters are called hyperbolic encounters. In encounters with low relative velocity, the trajectories are much more complex (see Fig. 2.11).

Let's do some simple math:

(1) The Jacobi constant of the asteroid is the same in A and B (because it is a law of conservation of the motion of one asteroid in the restricted model with Jupiter in circular orbit): $\mathcal{J}_A = \mathcal{J}_B$.
(2) The potential energy of gravitational attraction of the asteroid by the planet is the same in A and B (since the potential energy depends only on the distance and the points A and B are equidistant from the planet).

From these two relations plus the Jacobi integral, we obtain:

$$E_B - E_A = n(S_B - S_A), \tag{2.2}$$

that is, the variation of the energy of the asteroid is proportional to the difference between its angular momentum at the exit and entrance of the sphere of influence. If the motion of the asteroid were not deflected by the planet, the variation of both quantities would be zero. That is, $E_{B'}=E_A$ and $S_{B'} = S_A$.

The sequence of calculations, if done with precision, involves some complications. We will therefore make only a very approximate analysis. The angular momentum S at point A is given by $S = r_A v_A \sin\alpha$ where α is the angle formed by the Sun-asteroid radius vector (or by a unit vector $\boldsymbol{\rho}$ in the direction of the heliocentric radius vector) and the velocity of the asteroid. The quantities r and v at points A and B are not equal, but if the asteroid crosses the sphere of influence with great velocity, v_A and v_B are very close. (In fact, a good approximation is obtained by equalizing the relative velocities, i.e., $|\mathbf{v}_A - \mathbf{v}_0| \simeq |\mathbf{v}_B - \mathbf{v}_0|$.) The only parameters that are very different at points A and B are the angles α and β.

If the asteroid crosses the planet's orbit passing behind the planet (both orbits being assumed heliocentric and direct, that is, prograde) then, $\beta \gg \alpha$ (see Fig. 2.10, left) and therefore, $S_B >> S_A$. As a consequence $E_B > E_A$, that is, the heliocentric energy of the asteroid increases.

The same reasoning can be repeated if we change the configuration shown and consider that the asteroid crosses the planet's orbit in front of the planet. In this case, the result would be reversed, $\beta \ll \alpha$ (see Fig. 2.10, right) and $E_B < E_A$. The asteroid would have its heliocentric energy decreased.

In these two examples, we considered asteroids crossing the planetary orbit from inside to outside. They can be repeated with asteroids moving from outside to inside the planetary orbit, with the same results.

The asteroid has its energy increased when it passes behind the planet and decreased when it passes in front of it. The consequences of the energy change in a close encounter can be discussed by remembering some basic rules of Keplerian mechanics: (1) the energy of an asteroid in an elliptic orbit is negative and inversely proportional to its average distance from the Sun ($E = -GMm/2a$); (2) a zero or positive energy means that the asteroid has sufficient mechanical energy to escape from the Solar System; (3) $S < 0$ means retrograde orbit, that is, the motion on it is

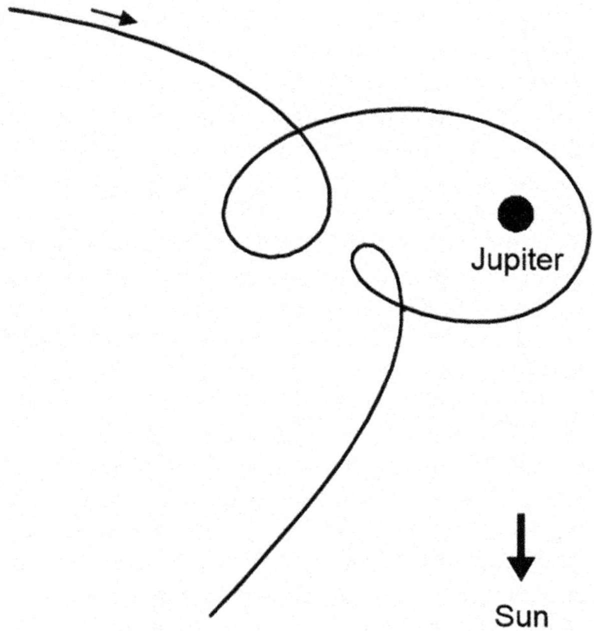

Fig. 2.11 Simulation of the close low-speed encounter of comet 65P/Gunn with Jupiter between 1868 and 1882. Minimum distance: 54 million km. The comet and Jupiter move around the Sun in counterclockwise orbits. The comet appears to be moving clockwise in the initial arc of the trajectory shown because, before the closest encounter, its motion around the Sun is slower than that of Jupiter (held fixed in this figure). Source: Istituto di Astrofisica Spaziale—CNR, Italy

in a direction opposite to the direction of the planetary motions. Some implications, by way of example: (1) An asteroid which crosses the orbit of a planet behind the planet has its energy and, therefore, its average distance from the Sun increased; if the passage is closer, the energy gain may be large and transfer the asteroid to a hyperbolic orbit and escape from the Solar System; (2) An asteroid that crosses the orbit of a planet in front of the planet has its energy, angular momentum, and the average distance of its orbit to the Sun decreased.

The two occurrences mentioned above are observed in many cometary orbits. Long-period comets, in general, have large eccentricities and cross the orbits of the planets with large velocities. With asteroids and short-period comets, the approach to Jupiter may occur with low relative velocity and have a very different geometry; the comet (or asteroid) may circle the planet and remain attached to it for a long time before leaving its neighborhood or even colliding with the planet. One example is the periodic comet 65P/Gunn, which has frequent close encounters with Jupiter. In the most spectacular of these encounters (Fig. 2.11), which lasted from 1872 until 1882, the comet remained all this time within the sphere of influence of Jupiter and decreased its average distance from the Sun from 6.4 to 4.1 AU (the orbital period decreased from 16.3 years to 8.4 years). The orbit that was larger than Jupiter's became smaller (Carusi et al. 1985).

A more complex example is the comet Shoemaker-Levy 9, whose primitive orbit could not be well determined. Besides a very close encounter with Jupiter, it may have also had another one with one of its larger satellites. It lost much energy and ended in a very eccentric orbit around Jupiter that led it to collide with the surface of the planet.

The asteroids are generally located in non-intersecting orbits internal to Jupiter's orbit. There is no simple figure capable of synthesizing what may happen in a close approach to Jupiter. If the encounter is very close, the exchange of energy and angular momentum with the planet may be large enough to transfer the asteroid to a new orbit of great eccentricity. Then it may cross the orbit of the inner planets, or even that of Jupiter, and become prone to a future close encounter able to eject the asteroid to the farthest reaches of the Solar System or put it in a collision course with the Sun. One consequence of this fact is that the main asteroid belt, where almost 95% of the known asteroids are found, ends far away from Jupiter's orbit (see Fig. 2.1).

Figure 2.10 shows that the most obvious change an asteroid (or comet) undergoes in a close encounter with a planet is that of its angular momentum. Therefore, we should expect that the eccentricity of the orbit also changes. The changes in eccentricity e and inclination i are related by a variant of the Jacobi integral called the Tisserand invariant, which is obtained from the Jacobi integral in the spatial case neglecting the term W_p (which is very small far from the disturbing planet), and writing the energy and angular momentum in terms of the asteroid's orbital elements (with S now standing for the angular momentum of the asteroid's motion projected onto the plane of the planet's motion):

$$\mathcal{T} = \frac{1}{\alpha} + 2\sqrt{\alpha(1-e^2)}\cos i = \text{const.} \tag{2.3}$$

where α is the ratio between the average distances from the Sun to the orbits of the comet (or asteroid) and of the planet (Dvorak and Lhotka 2013). One criterion was introduced by the French astronomer Félix Tisserand, as an attempt to identify comets observed at different epochs in different orbits. Since the value of $\mathcal{T}$ is not altered by a close encounter, if two cometary orbits determined at different times have the same $\mathcal{T}$ value, one can assume that they belong to the same comet and are different just because of a close encounter with Jupiter between the two determinations. Precise simulations of the comet's motion can then confirm the reality of this fact.

2.4 Asteroids in the Restricted Three-Body Elliptic Model

Consider again a simple planar model with Jupiter on a fixed orbit and the asteroid moving in the plane of Jupiter's orbit. But let us now consider Jupiter in an elliptical orbit. The simplest reference system that we can devise in this case has the center at the Sun and the x-axis oriented in the direction of the perihelion of Jupiter's orbit (Fig. 2.12). To characterize the configuration of the system at a given moment, we need 3 coordinates: The x, y coordinates of the position occupied by the asteroid and

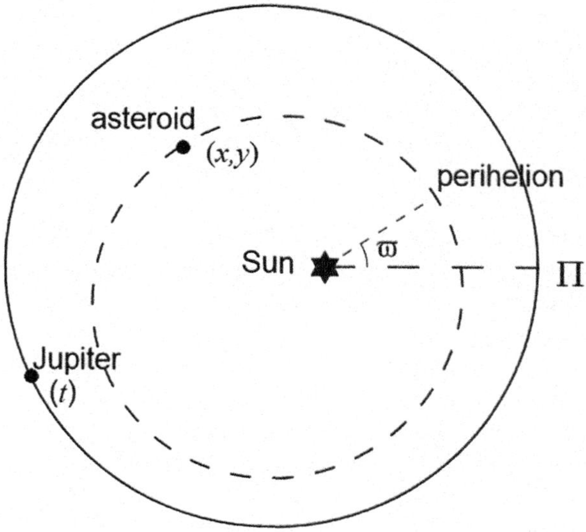

Fig. 2.12 Asteroid and Jupiter. The elliptical orbit of Jupiter is fixed. Π is Jupiter's perihelion. The asteroid's orbit is variable

an angle allowing us to know at which point of its orbit is Jupiter. Geometrically, the system has two degrees of freedom (since the only unknown variables are x and y). But the configuration of the system is only complete when the position of Jupiter is also given. The motion of Jupiter is known, but the position of Jupiter depends on time. Therefore, we need an additional parameter that can be time itself, and this means increasing one degree of freedom. Then, the restricted elliptic three-body problem has three degrees of freedom, and we cannot apply to it the same techniques that we have used in the case where Jupiter was considered to be in a circular orbit (in which case the system had only 2 degrees of freedom).

The position of the asteroid is fixed by the coordinates x and y, and its motion by the momenta (quantities of motion) $p_x = m\dot{x}$ and $p_y = m\dot{y}$. These four quantities can be replaced by another group of four quantities being 2 angles and 2 momenta that can be defined using, at each instant of time, the instantaneous orbit of the asteroid (because of the perturbation due to Jupiter, the orbit of the asteroid is not a Keplerian ellipse, but can be assimilated to a Keplerian ellipse that is continuously being deformed). The size and shape of the ellipse are given, at each instant of time, by the semi-major axis a and the eccentricity e, and the orientation of the ellipse by the longitude of the perihelion ϖ. The position of the asteroid on the ellipse is given by its mean longitude[4] λ. The two angles ϖ and λ can be used as new coordinates defining the position of the asteroid, and the two new momenta will be functions of a and e. The variables of the restricted elliptic problem are completed with the mean longitude λ' of Jupiter (which is a linear function of time).

[4] The mean longitude is not a geometrically defined angle. It measures the fraction of the period elapsed since the body passed through the origin axis (in Fig. 2.12, the origin axis is the horizontal dashed line indicating the perihelion of Jupiter's orbit).

Let us now consider the proper frequencies of this system: n, n', and g. The frequency $n = 2\pi/T_{\text{ast}}$ is the average angular velocity of the asteroid's motion, $n' = 2\pi/T_{\text{Jup}}$ is the average angular velocity of Jupiter's motion, and g is the velocity of the angle ϖ. If the motion of the asteroid were not being disturbed by Jupiter, the angle ϖ would be constant. But the gravitational attraction of the asteroid by Jupiter causes this angle to vary. Suppose now that the asteroid is resonant, and let us introduce the commensurability relation between the asteroid period and Jupiter's period. Let us say, for instance, that the quotient of the two periods is: $T_{\text{Jup}}/T_{\text{ast}} \sim p/q$ where p, q are two integers. For example, in the cases discussed in Sects. 2.2.1, 2.2.2, and 2.2.3, these rational numbers are 2/1, 3/2 and 3/1, respectively.

In 12 years (approximate period of Jupiter), the angle λ' completes one revolution. Similarly, the angle λ completes one revolution every $\sim 12(q/p)$ years (i.e., in the cited examples, approximately 6, 8, and 4 years, respectively). In these cases, the angles λ and λ' are strongly correlated.

The longitude λ is a rapidly changing angle. The effects of the gravitational attraction of Jupiter are slow and much smaller. To highlight them, it is convenient to replace the angle λ with the combination $\psi = q\lambda - p\lambda'$ in which the two longitudes appear multiplied by the integers q and p in such a way that the two parts compensate one another. Then, only one fast variable will remain in the problem. Actually, to preserve the symmetries existing in the differential equations, we adopt the angles ψ, $\theta = \psi + (p-q)\varpi$, and $Q = \lambda - \lambda'$. The first two are slowly varying, and the third, subtraction of the two mean longitudes, is fast (i.e., of short period).

The restricted circular model of Sect. 2.2 had 4 dimensions and an energy integral that allowed us to reduce the range of motion to a space with 3 dimensions. This space was then cut by a plane, on which we were able to see the intersections of the motion. Now the situation is different. If we proceed in the same way, we will be cutting with a plane not a space of 3 dimensions, but a space of 5 dimensions. After the cut, each intersection will have 4 dimensions and a planar figure is insufficient to represent it. Even if the motion is regular, the representation on a plane is a projection and mix different parts of the figure. We have an excess of 2 dimensions, and the figures on a plane become difficult to interpret. Although sometimes this resource is used in problems with more than 2 degrees of freedom, it is important to note that the planar figures thus obtained are no longer Poincaré maps. In Poincaré maps, each point in the plane corresponds to a single solution; in the plane figures obtained from systems with more than 2 degrees of freedom, each point may represent an infinity of solutions.

2.5 Reduction to Two Degrees of Freedom. Averaging Processes

To study a problem of this type—and they are abundant in Physics and Astronomy—in which slow variables and fast variables appear simultaneously, we use averaging processes to eliminate the fast variables from the problem. Let's consider a simple example. Let's represent the coordinates generically by $\mathbf{x}$ and the momenta generically by $\mathbf{p}$. The equations are $\dot{\mathbf{x}} = \mathbf{f}(\mathbf{x}, \mathbf{p}, t)$ and $\dot{\mathbf{p}} = \mathbf{g}(\mathbf{x}, \mathbf{p}, t)$.

Since the coordinates adopted are angles, the functions that appear in the right-hand side of the equations are, in general, series of trigonometric terms (Fourier series), for example

$$\sum A_{\mathrm{j}}(\mathbf{p}) \cos(j_1\psi + j_2\theta + j_3 Q).$$

A very popular method to obtain average equations is the scissors method. We take the Fourier series representing the right-hand sides of the differential equations and cut out all the terms that have in their arguments the fast angle(s). In the case considered in the previous section, for instance, we eliminate all terms in which the angle $Q = \lambda - \lambda'$ occurs. This operation reduces the problem to two degrees of freedom, since now the right-hand sides depend on only two coordinates: ψ and θ. Moreover, in systems of this type, if the angle Q disappears from the right-hand sides, the momentum associated with Q becomes constant. Since the angle Q disappeared from the equations, it is no longer necessary to solve the equation dQ/dt at the same time as the others. It can be studied after the solution for the rest of the system is known. The right-hand sides of the other equations still depend on the angular momentum associated with Q, but it is now a constant and can be considered only as a known constant parameter in the other equations. This leaves in the equations only the other two angles, which are slow.

This provides a way to look at the long-term evolution of the process. The new system evolves slowly. After eliminating the rapid fluctuations, we are left with the long-term evolution of the system. The simple procedure described has, in fact, limitations. But the corrections that should be introduced to validate the averaging are very well-known (Ferraz-Mello 2007) and, if necessary, it is possible to perform the averaging rigorously. We know how to construct a transformation of variables $(\mathbf{x}, \mathbf{p}) \Rightarrow (\mathbf{x}^*, \mathbf{p}^*)$ such that the rapid changes are no longer present in the transformed equations. This operation has an error that depends on the order of approximation used in the averaging process. In the most summary process, which we called scissors method, the accuracy of the average has the same order of magnitude as the perturbations of the motion of the asteroid due to Jupiter, i.e., the ratio of the mass of Jupiter to the mass of the Sun (approximately 0.001). However, using a properly constructed transformation, this accuracy can be considerably improved.

In general, to know the main characteristics of the motion of the asteroids, the approximate process is sufficient. We are solving a physical science problem, not a mathematical problem. In general, the limitations of the models that we adopt introduce into the equations of motion errors or inaccuracies as important, or more important, than the mathematical inaccuracies of the processes used to solve them.

2.5.1 From "Himmelsmechanik" to "Atommechanik"

The average methods of perturbation theory appear in Celestial Mechanics connected with the names of Charles-Eugene Delaunay, Karl Bohlin, Henri Poincaré, Hugo von Zeipel, and Gen-Ichiro Hori (sometimes Hori-Lie, as Hori uses the Sophus Lie series to perform this elimination). In Physics, the classical methods studied by Poincaré (which he called Lindstedt's methods) appear for the first time in Max Born's "Atommechanik" (Born 1925). In the preface to that book, Max Born wrote:

> The title Atommechanik given to these lectures, which I delivered in Göttingen during the session 1923–24, was chosen to correspond to the designation Himmelsmechanik (i.e., Celestial Mechanics). As the latter name covers that branch of theoretical Astronomy which deals with the calculation of the orbits of celestial bodies according to the laws of Mechanics, so the name Atommechanik is chosen to signify that the facts of Atomic Phyics are to be treated here with special reference to the underlying principles of Mechanics.

It is interesting to observe that the great source of knowledge in Mechanics that inspired the principles of the Atomic Mechanics of Johannes Burgers, Karl Schwarzschild, Arnold Sommerfeld, Paul Epstein and Max Born was Celestial Mechanics, and one of the main bibliographic references throughout this period, for reasons of idiom but also of merit, was "Die Mechanik des Himmels" by Carl V. L. Charlier. It was in this book that physicists got the idea of the action-angle variables which played a fundamental role in the establishment of the "Old Quantum Theory" (prior to Quantum Mechanics), and the modern methods of averaging (or perturbation calculations, as they are also called), which are successfully used until today.

2.6 The Alinda gap

The Alinda gap is the one located in the 3:1 resonance with Jupiter (see Fig. 2.2), where the asteroids have period close to 3.95 years (i.e., about 4 years, one third of the Jupiter period) (see Fig.2.8). The name comes from (887) Alinda, discovered in 1918 by Max Wolf, the first asteroid discovered moving within the resonance. Some ancient texts refer to these asteroids as Hestia-type asteroids. The asteroid (46) Hestia, discovered in 1857 by Norman Pogson, was the first of a group close to the 3:1 resonance (but outside of it).

The simplest model in which the asteroid's motion around the Sun is only perturbed by Jupiter moving in a fixed elliptical orbit is enough to study the motion of these asteroids. In this case, the asteroid makes three revolutions around the Sun while Jupiter makes one. We have the period ratio $T_{\rm Jup}/T_{\rm ast} \sim 3/1$.

The 3 angular coordinates best suited to describe this motion of 3 degrees of freedom, chosen so as to preserve the symmetries of the equations of motion, are:

$$\begin{aligned} \psi &= \lambda - 3\lambda', \\ \theta &= \psi + 2\varpi, \\ Q &= \lambda - \lambda'. \end{aligned} \tag{2.4}$$

(see Sect. 2.5). The angle Q varies rapidly, and to reduce the system to only 2 degrees of freedom, it must be eliminated using an averaging method—the scissors method is sufficient. The angles ψ and θ vary very slowly. If the asteroid is "inside" the resonance, these angles only oscillate with oscillation periods of about 300 years. The difference between ψ and θ is the angle 2ϖ which has a much slower motion.

To construct the Surface of Section, we look for a plane that is repeatedly crossed by the solution. One can use for this purpose, for instance, the angle θ. At each oscillation of the angle θ, the system crosses the plane corresponding to its mean value, $\theta = 180°$ (see Sect. 2.2.3). Also, to avoid the problems discussed in Sect. 1.12, we do not consider all the intersections of the trajectories with the plane, but only those in which the plane is crossed in a fixed direction (e.g., $\dot{\theta} < 0$).

A good representation of the intersections of the trajectory with the plane $\theta = 180°$ is obtained using polar coordinates: the radius vector is the eccentricity of the orbit at the moment of intersection, and the polar angle is the longitude of the perihelion (referred to the fixed perihelion of Jupiter) at that moment. Their equivalent rectangular coordinates are $e.\cos(\varpi - \varpi_{\mathrm{Jup}})$, $e.\sin(\varpi - \varpi_{\mathrm{Jup}})$ (Fig. 2.13).

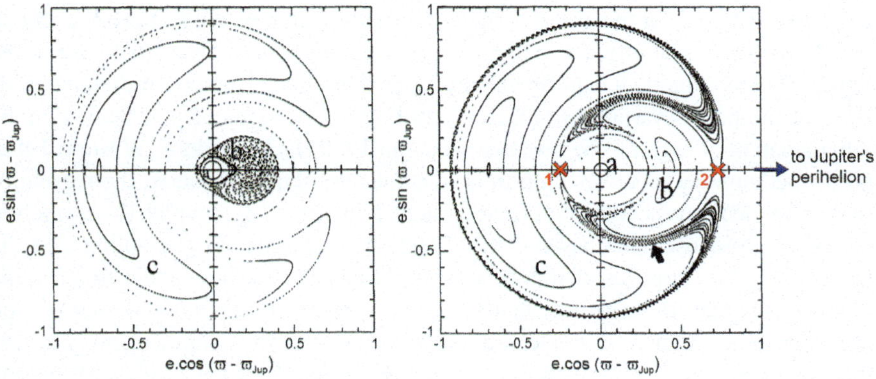

Fig. 2.13 Surfaces of Section of the motion of an asteroid in the 3:1 resonance at two different energy levels. a, b, c indicate three different regimes of motion around 3 centers corresponding to 3 stable periodic orbits. The 2 crosses indicate saddles corresponding to 2 unstable periodic orbits. The arrow indicates the bridge connecting the neighborhoods of the two saddles. e is the eccentricity of the asteroid orbit and ϖ, ϖ_{Jup} are the longitudes of the asteroid and Jupiter perihelions. We adopt $\varpi_{\mathrm{Jup}} = 0$ (Reproduced from Ferraz-Mello et al. (1996). Copyright 1996 by Springer)

The set of figures thus obtained is a tomography of the phase space in which each figure corresponds to a given level of energy and each curve to a different initial value of the angular momentum. The paths followed by the asteroids were simulated by means of numerical integrations. These computations are made with high precision. The chaoticity of most of the examples shown corresponds to the tenfold increase in the errors (loss of a digit in the information) in times of the order of a few tens of thousand years; computations simulating the trajectories for a few hundred thousand years can be done with enormous accuracy.

Figure 2.13 shows two Surfaces of Sections of the motion of an asteroid in 3:1 resonance corresponding to two different energy levels and exhibiting three different regimes of motion indicated by a, b, c. In the first section, the resonance center lies at a low eccentricity and acts mainly on regime b; in the second section, of higher-energy, the resonance center lies at a larger eccentricity, acting on regimes b and c and creating a confluence common to the 3 regimes.

In these figures, the times required to perform a complete cycle in regime b are between 30 and 60 thousand years, and in regime c, about 200 thousand years. There is no effect on the results shown that can be attributed to the propagation of computational errors. They are only consequences of the chaotic evolution of the system.

2.6.1 Regimes of Motion in the Resonance 3:1

The resonance 3:1 is the best known of the many asteroidal resonances. Until 1982, the only known mode of motion in this resonance was the ordinary regime (marked with a in Fig. 2.13), which occurs at low eccentricities around a center close to the origin. In this regime, the eccentricity is almost constant (it has only a small periodic variation) and the perihelion retrogrades (it moves clockwise in the figure). Some numerical experiments by Hans Scholl and Claude Froeschlé (1974) showed anomalous variations of the eccentricity, but the dynamics of motion on the Alinda gap only began to be revealed when, in 1982, Jack Wisdom (Wisdom 1982) performed simulations over long time intervals and thus discovered the regime of motion of medium eccentricities (marked with b in Fig. 2.13) in which the perihelion of the asteroid oscillates around the direction of the Jupiter perihelion and the eccentricity has large oscillations, reaching almost 0.4 at the outer edge of the domain in which this regime occurs. As the system of equations of this motion is non-integrable, there is no separation between the two regimes of motion, and the asymptotic branches coming out of the saddle points of the Surface of Section become entangled (see Figs. 1.14 and 1.15).[5]

[5] For some values of the initial energy, the entanglement formed at the boundary of the two regimes almost completely fills the region of medium-eccentricity motions.

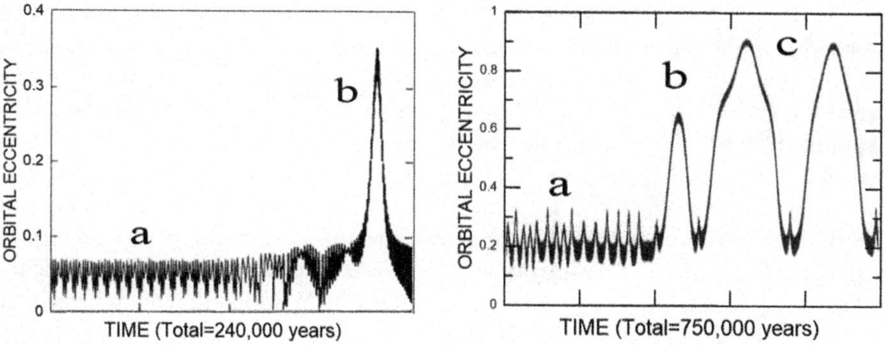

Fig. 2.14 *Left*: Evolution of the eccentricity in a solution in which the motion starts in regime a and transitions to regime b. (Adapted from Wisdom (1983). Copyright 1983 by Elsevier.) *Right*: Evolution of eccentricity in a solution in which the motion starts in the regime a, transitions to regime b, and reaches regime c (Adapted from Ferraz-Mello and Klafke (1991). Copyright 1991 by Springer)

The boundary between the two regimes is permeable, and a solution can pass through it and change from one regime of motion to another. In an example calculated by Wisdom (first panel of Fig. 2.14), the motion starts with the asteroid somewhere in region (a); it circulates in the inner part for long time in an apparently regular way with the eccentricity oscillating almost periodically. The oscillation patterns repeat regularly, almost identically, and a spectral analysis of this part of the motion will probably show some peaks corresponding to well-defined frequencies. Then, suddenly, the pattern begins to change until the eccentricity jumps as the orbit invades region (b). This transition can happen because the system of equations of the motion of the asteroid is not integrable. Initial conditions not very far from the boundary between the two regimes can lead to solutions whose regime alternates between them in a way that cannot be predicted for long times.

However, the models used until then could not be used to study what happens at large eccentricities; strictly speaking, their limit of validity was just below $e = 0.4$ (see Ferraz-Mello 1994c). The results obtained for the transition zone between a and b are valid, but the outermost part of region b could not be studied.

A third regime of very high eccentricities, also called corotation, was discovered in 1991 (Ferraz-Mello and Klafke 1991; Ferraz-Mello et al. 1993), thanks to the construction of semi-analytical models valid for high eccentricities. In this third regime (marked with c in Fig. 2.13), the perihelion has oscillations that can be huge (almost 180°), around the direction opposite to that of Jupiter's perihelion, and the eccentricity can get very close to 1. This third regime is well separated from the inner ones by apparently regular motions (closed curves enclosing the chaotic region in the first panel of Fig. 2.13), but for some values of the initial energy, these apparently regular orbits cease to exist, and asymptotic branches appear that leave the neighborhood of one saddle point and move toward the other (marked with X in the second panel of Fig. 2.13). These branches form a bridge connecting the neighborhood of one saddle to another (indicated with an arrow in the figure). In this case, we have the so-called heteroclinic points where stable and unstable branches coming from two different saddles intersect.

The *heteroclinic bridge* formed by the tangle of asymptotic branches emanating from the saddle points makes the various regimes of motion permeable, and the intermittencies involving this bridge allow trajectories beginning with eccentricity close to 0.2 to circulate for a long time in the low-eccentricity domain and, suddenly, jump to higher eccentricities. Then, as predicted by Wisdom, they may invade the region of medium eccentricities (b) or, occasionally, enter the heteroclinic bridge and reach the high-eccentricity domain (c).

The second panel of the Fig. 2.14 shows an example calculated by Sylvio Ferraz-Mello and Júlio César Klafke (1991). In this example, the motion starts with the asteroid circulating inside the low-eccentricity region (a). After some time, the pattern changes, and the solution leaps; it makes one cycle in the region (b) and then transitions to region (c). The maximum value of the eccentricity reached in this example is 0.9.

2.6.2 The Origin of the Alinda Gap

From the standpoint of asteroid system architecture, the important fact is that when the eccentricity of an asteroid in this group is close to 0.4, the minimum distance of the asteroid from the Sun is $\sim$1.3 AU. This is smaller than the average distance of Mars from the Sun. Then, the asteroid's orbit crosses the orbit of Mars allowing close encounters between the asteroid and the planet to occur. Similarly, when the eccentricity reaches 0.9, in addition to the possibility of crossing the orbit of Mars, we have an orbit crossing the orbits of the Earth, Venus, and even Mercury.

The resonance 3:1 is very close to Mars. As the asteroid's eccentricity increases, its orbit will cross Mars' orbit. It can then have a close encounter with Mars (without colliding) and exchange energy with the planet, as was discussed in Sect. 2.3. If the energy that it gives or receives is large enough, the orbital period of the asteroid is changed and ceases to be $\sim 1/3$ of the period of Jupiter.

The transition to regime b, discovered by Wisdom, was the first explanation for the almost non-existence of asteroids in the Alinda gap. But Earth and Venus are planets of masses 10 times that of Mars. Therefore, a close encounter with either Earth or Venus is much more efficient at extracting the asteroid from the resonance than an encounter with Mars. But the Surfaces of Section show that, in general, bundles of closed curves exist involving domains a and b (see the first panel of Fig. 2.13) so that the chaos seen in these domains is confined and does not communicate with the region of high eccentricity; the heteroclinic bridge that allows access to high eccentricities (indicated with an arrow in the second panel of Fig. 2.13) exists only in a very narrow energy range.

The model that includes only the Sun, the asteroid, and Jupiter (in a fixed orbit) and places them all in the same plane can yet be improved. In a model that also includes the outer planets Saturn, Uranus, and Neptune, the heteroclinic bridge widens, allowing an asteroid placed in the interior of the resonance more often to come close to the Earth and Venus. Alessandro Morbidelli and Michelle Moons (1995) showed that in a more realistic model, the fate of an asteroid whose period is 1/3 of Jupiter's period is mainly regulated by the transitions to the regime of corotation. Although the transitions to this regime are less frequent than transitions to the medium-eccentricity regime, they are fatal. Mars has a very small mass, and an asteroid can remain in the medium-eccentricity regime for millions and millions of years without getting close enough to Mars to be deflected from its orbit. But in the corotation regime, the orbit becomes so elongated, and for such long times, that an approach to Earth or Venus capable of pulling the asteroid off the Alinda gap has a better chance of occurring. Reality can be even more violent, with some orbits becoming so eccentric that the asteroid collides with the surface of the Sun (Farinella et al. 1994; Ferraz-Mello 1999; Gladman et al. 1997). In these extreme cases, the asteroid will be broken into pieces by the tidal stretching of its body caused by the Sun and destroyed by the extreme heating, even if the eccentricity is not so high as to cause it to fall on the surface of the Sun.

The above-discussed effects can be verified by observing the distribution of the semi-major axes and eccentricities of the asteroids (Figs. 2.15 and 2.16). Figure 2.15 shows about 1000 bright asteroids located in the vicinity of the resonance 3:1. The figure clearly shows the gap in resonance 3:1 and the absence of bright asteroids in the interval between 2.498 and 2.512 AU (the center of the exact resonance is at 2.500 AU). The asteroids shown outside the gap are asteroids very close to the resonance but their periods are smaller (on the left-hand side) or larger (on

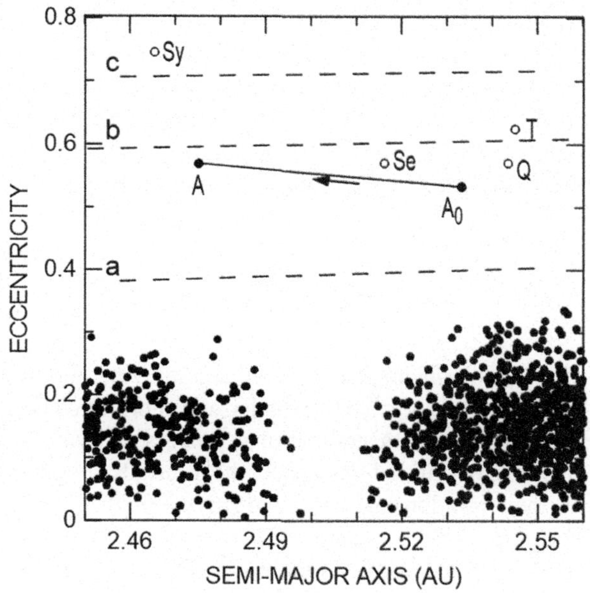

Fig. 2.15 Distribution of semi-major axis and eccentricity of the asteroids of the resonance 3:1 with absolute magnitude up to 14. A and A_0 indicate the asteroid (887) Alinda now and at the time of its discovery, one century ago. Also included are some asteroids of smaller magnitude: (1915) Quetzalcoatl, (2608) Seneca, (3360) Syrinx, and (4179) Toutatis. The lines a, b, c indicate the lower limits from which the orbit of an asteroid in this resonance crosses the orbits of Mars, Earth, and Venus, respectively

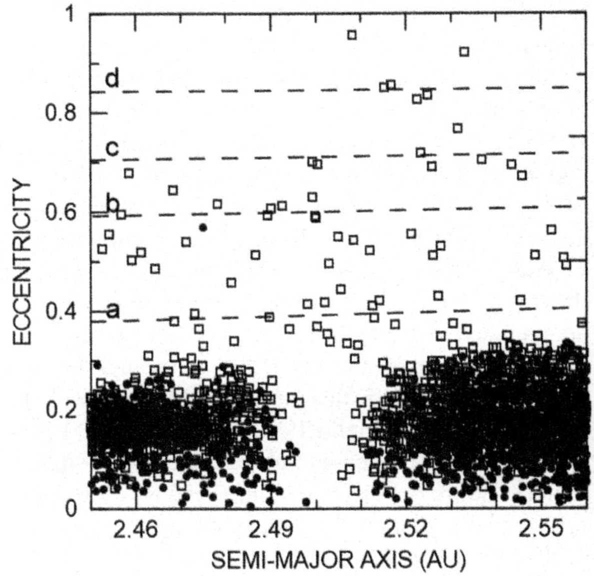

Fig. 2.16 Distribution of semi-major axis and eccentricity of the asteroids of the resonance 3:1 with absolute magnitude greater than 18 (squares), compared to that of the bright asteroids shown in Fig. 2.15 (dots). The lines a, b, c, and d indicate the lower limits from which the asteroid orbits cross the orbits of Mars, Earth, Venus, and Mercury, respectively

the right-hand side) than ~3.95 years. At high eccentricities, we see the asteroids (887) Alinda (currently and at the time of its discovery, one century ago) and the less bright asteroids (1915) Quetzalcoatl, (2608) Seneca, (3360) Syrinx, and (4179) Toutatis, included because they have been the subject of several detailed dynamical studies (see Sect. 2.6.4). These asteroids are moving in the corotation domain C and cross the orbit of several of the inner planets. The asteroid (887) Alinda has been observed continuously since its discovery one century ago, allowing the observational confirmation of its libration around the central position of the resonance. Points A and A_0 represent its semimajor axis and eccentricity, currently and shortly after its discovery (Berliner Astronomisches Jahrbuch 1919; Ferraz-Mello 1989), respectively. The segment connecting them shows the displacement due to the asteroid libration between the two epochs.

2.6.3 *Near-Earth Asteroids (NEAs)*

The chaotic phenomena that we are studying are purely gravitational. The fainter asteroids are, in general, smaller and more affected by non-gravitational accelerations unrelated to gravitational processes. Heated by sunlight, when the stored energy is re-emitted into space, the asteroids are pushed in a direction that depends on their rotation and the conductivity of their surface. This effect is known as Yarkovsky effect. In large asteroids, the resulting acceleration is negligible, but in small asteroids, it affects their average distance from the Sun. Measured variations of this distance for the asteroids of the Karin family can reach 15 m/year in the case of an asteroid with 2 km of diameter (Nesvorný and Bottke 2004).

The limit adopted in Fig. 2.15 excludes tens of thousands of small asteroids that have been discovered in the last decades. The ideal would have been to set a size limit, but only few asteroids have had their diameters measured. The absolute magnitude H of an asteroid is related to its diameter by an approximate logarithmic law:

$$H = 12.9 - 4.45 \log_{10} \left(\frac{D_{\mathrm{km}}}{10} \right) + \delta H \tag{2.5}$$

where δH is a quantity that depends on the nature of the asteroid, varying approximately between -1.5 and $+1.5$ depending on whether it is an asteroid whose surface is more reflective or more dull. By setting as a limit, the absolute magnitude $H = 14$, we included in Fig. 2.15 all asteroids near the resonance 3:1 whose diameter is greater than 10 km. Because of the lack of knowledge of the corrective term δH for most asteroids, many of the included asteroids are smaller than 10 km, and have their brightness increased not because they are larger, but because of a higher reflectivity of their surfaces. Without specific measurements, it is not possible to differentiate them, so we had to leave them in that figure.

The smaller asteroids have their dynamics governed by both the resonance and the Yarkovsky effect. This can be seen in Fig. 2.16 where, besides the brightest asteroids, we did include those of absolute magnitude $H > 18$. This limit corresponds, in general, to sub-kilometric asteroids (with, at most, D = 2 km). Two facts in this figure call our attention. The first is that, at low eccentricities, the gap is preserved. The time required for an asteroid 2 km in diameter to cross the gap is very large. It is larger than 100 million years, that is, 2 or 3 orders of magnitude longer than the typical times for the manifestation of chaotic transitions. Thus, at low eccentricities, we see many sub-km asteroids near the edge of the gap but none in its interior. The second is the large number of high and very high eccentricity asteroids crossing the orbits of the inner planets.

Many of them cross the Earth's orbit and are part of the population of Near-Earth Asteroids (NEAs) that has been exhaustively studied in the past 30 years, given the perception that they can collide with the Earth. Asteroidal resonances are inexhaustible sources of NEAs. In the case of resonance 3:1, the mechanism can be inferred from Fig. 2.16. Small asteroids, driven by the Yarkovsky effect, invade the edges of the Alinda gap and are thus subjected to chaotic transitions, which cause their orbital eccentricities to increase significantly, allowing them to reach the orbit of the Earth.

This process, visualized at the end of the last century, was very well described in a cartoon by the astronomer Enzo Zappalà (Fig. 2.17), showing the various stages of the evolution of NEAs, from the collision and fragmentation of larger asteroids close to the resonance until their arrival to the neighborhood of the Earth.

Fig. 2.17 Formation and evolution of NEAs. © Enzo Zappalà (Authorized reproduction)

2.6.4 Alinda, Quetzalcoatl, Seneca, Syrinx and Toutatis

The Alinda gap is a region of strongly chaotic motions and an asteroid placed there is bound to have a close encounter with one of the inner planets, or with the Sun, and leave the resonance. The asteroids (887) Alinda, (1915) Quetzalcoatl, (2608) Seneca, (3360) Syrinx, and (4179) Toutatis (studied by Andrea Milani and collaborators as part of the Spaceguard project (Milani et al. 1989)), and many other small asteroids are in this region. But they are not permanent asteroids of this region. They have been on the Alinda gap for a very short time. They are asteroids that were thrown into this region by very recent events, which changed their energies and allowed them to be captured by the resonance. For example, Alinda was captured in the resonance about 60 thousand years ago (Milani et al. 1989), coming from a high eccentricity orbit, not far from the resonance 3:1 (Fig. 2.18). These asteroids are very small and are most likely fragments whose largest diameter does not reach 10 km (Quetzalcoatl may be much smaller than 1 km). Alinda, Quetzalcoatl, and Seneca have rather eccentric orbits (~0.57–0.58) but do not reach the limit of 0.60 necessary for them to cross the orbit of the Earth. But, sooner or later, all of them will have passages very close to the inner planets, and in these passages will suffer important variations in their energies and be expelled from the resonance to other regions in the asteroid belt or, if the encounter is very close, even out of the Solar System.

Toutatis (Fig. 2.19) is an eloquent example of an asteroid that is moving as expected. Its orbit is making one incursion through high eccentricities, as shown in the figures. It is currently at an eccentricity 0.62, and its perihelion lies inside Earth's orbit. It crosses the Earth's orbit every 4 years (twice, once coming in and

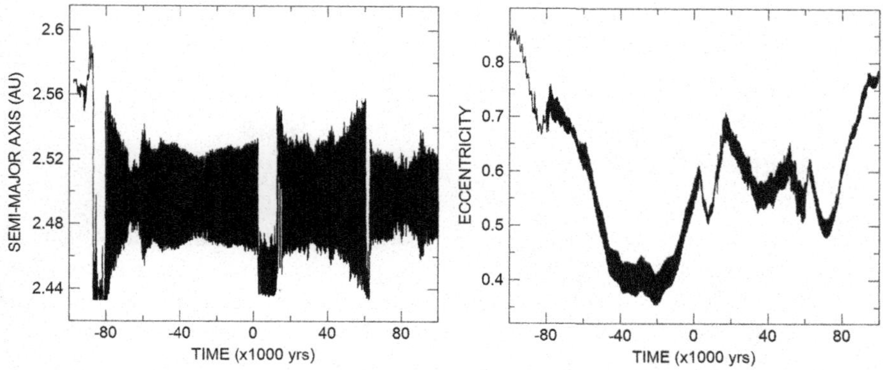

Fig. 2.18 Evolution of semi-major axis and eccentricity of (887) Alinda over 200 thousand years centered at the present time (Reproduced from Milani et al. (1989). Copyright 1989 by Elsevier)

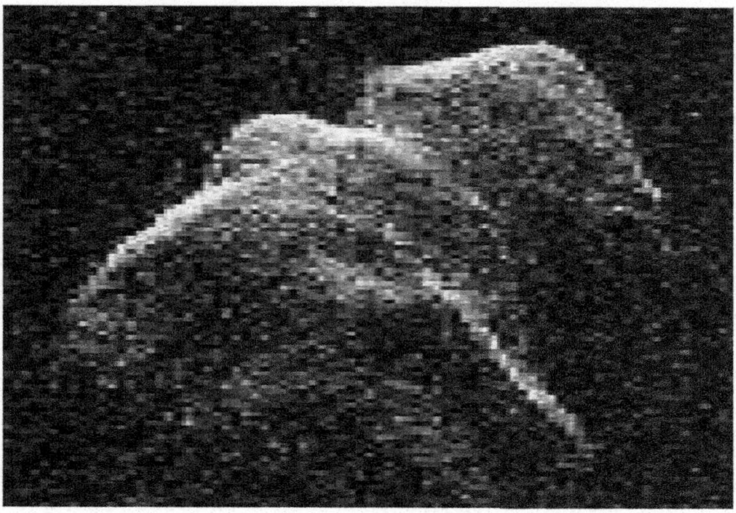

Fig. 2.19 Radar image of Toutatis on the near-Earth passage of 1996 (Credit: NASA/JPL)

once going out). Occasionally, it may cross the Earth's orbit at a time when the Earth is close to the point of intersection. For example, on August 12th, 1992, it passed 3.6 million km from the Earth, a distance less than 19 times the distance of the Moon. This proximity allowed it to be mapped with radar echoes, notwithstanding its small size ($1.9 \times 2.4 \times 4.6$ km). It will continue to have many close encounters with Earth in the future, escaping the resonance, and later (540 thousand years), will be ejected from the Solar System (Fig. 2.18).

Other asteroids have even higher eccentricities. (3360) Syrinx (which appears in the Spaceguard project documents under its former designation 1981 VA) has a higher eccentricity ($\sim$0.75), which allows it to reach the orbit of Venus! Some of the sub-km asteroids included in Fig. 2.16 have even larger eccentricities having their perihelions within 10 million kilometers of the Sun.

2.7 Digital Filtering

Averaging methods can quickly lead to large algebraic computations that are difficult to perform even with the help of programs allowing computers to do them. Moreover, they depend on the representation of the perturbing forces in Taylor series, which converge only for relatively low eccentricities. For instance, in the case of the asteroids of the resonance 2:1, the series converge only if the eccentricity is smaller than 0.2 (Ferraz-Mello 1994c). In many problems, the digital filtering method discussed below can perform numerical operations whose result is equivalent to an averaging.

Let us begin by analyzing the result of the numerical integration of the equations of an asteroid in the resonance 2:1. Let us take, for instance, the values of the semi-major axis of the asteroid calculated over 40,000 years. The solution (Fig. 2.20a) clearly shows a periodic variation with a period of almost 300 years, overlaid by a very rapid fluctuation due to the fast angle $(\lambda - \lambda')$.

Some techniques allow us to analyze the periodicities present in the solution. The best known of them is the Fourier transform. We build, with numerical integration, a table giving the orbital parameters at moments separated by equal time intervals. For example, for asteroids, we can store the values of the orbital elements every 20 or 30 days. Once we have built a table over a very long time, we can, by means of the Fourier transform, convert the temporal information contained in the table into information about the frequencies present in it. Doing this with the values shown in the figure, we obtain its spectrum (Fig. 2.20b). In this figure, the horizontal axis is the frequency axis, and the vertical axis gives the intensity with which the frequencies appear in the table data. The figure shows that there is a dominant line corresponding to the main visible oscillation of the order of 300 years, several peaks at low frequencies corresponding to variations whose periods range from 100 to thousands of years, and a large number of less well defined but still important lines corresponding to short period components (below 30 years). We also observe the clear separation between the lines corresponding to the high and low frequencies. This separation allows us to eliminate short-period (i.e., high-frequency) terms from the data. To do this, a fairly simple procedure is applied. We multiply the spectrum by a filter whose response is equal to 1 for the low-frequency part and 0 for the high-frequency part (Fig. 2.20c). The new spectrum is the product of the old spectrum and the filter response curve; it no longer has the high-frequency (short-period) terms (Fig. 2.20d). Now we have to perform the inverse operation of the Fourier transform, which allows the reconversion of the frequency spectrum into a time evolution table. The result is a function that is the convolution of the given function with the low-pass filter (Fig. 2.20e). It is clearly seen that the new function no longer has the high-frequency terms (as it was obtained by the inverse Fourier transform

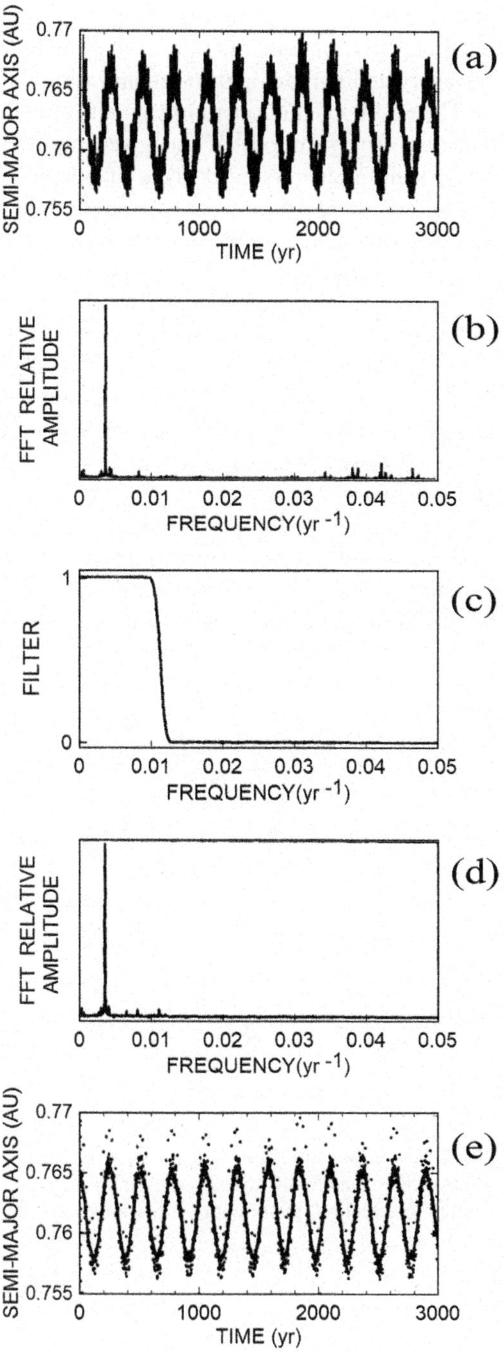

Fig. 2.20 Digital filtering. Example of the operation of a low-pass filter. (**a**) Time variation of the semi-major axis; (**b**) Fourier Transform (FFT) of the semi-major axis (frequency in cycles per year); (**c**) Frequency response of the low-pass filter; (**d**) Fourier Transform (FFT) of the semi-major axis after filtering; (**e**) Time variation of the semi-major axis after filtering (solid line) (Reprinted with permission from Michtchenko and Ferraz-Mello (1995) ©ESO)

of a spectrum that had no high-frequency terms). The effect of the convolution of the time variation with the filter is to eliminate the high frequencies. The signal was reconstructed without high-frequency terms (Ferraz-Mello 1994b).

The result of this operation is equivalent to the elimination of one degree of freedom. The solutions of the given problem, with 3 degrees of freedom, were numerically reduced to 2 degrees of freedom. The results in the filtered tables can be studied in the same way as we studied the restricted circular problem and other systems with 2 degrees of freedom: by constructing Surfaces of Section. The major difficulty is that it is not simple to obtain the corresponding transformation of the law of conservation of energy. That law is necessary to identify solutions of the same energy and construct maps showing the topological features of the system.

To prove that this numerical process really eliminated one degree of freedom of the system, it is enough to choose a domain of the given system, which is regular, and plot a solution in its Surface of Section. The result should be a well-defined curve, as shown in Fig. 2.21, which, even when magnified 40 times, still clearly shows a line.

The process of eliminating one degree of freedom by digital filtering consumes a large amount of computation time since we have to perform the integration of the complete equations, the Fourier transform operations for each coordinate, and the reconstitution of the filtered values by the inverse transform.

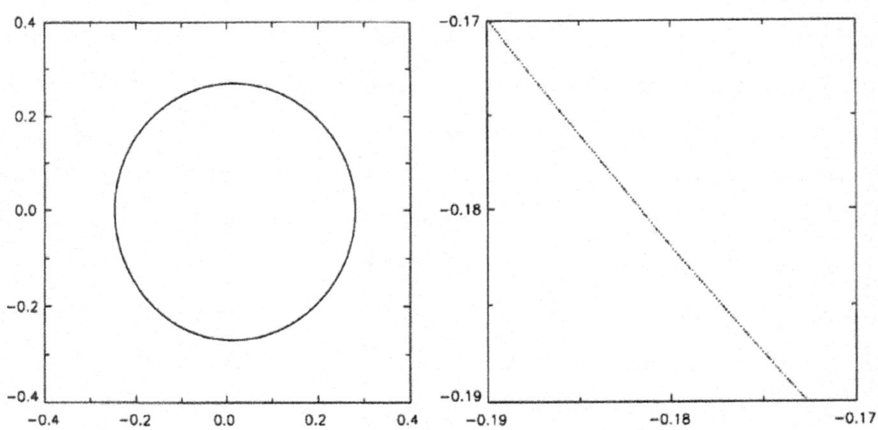

Fig. 2.21 Surface of Section constructed after the digital filtering. Left: Consequents of a regular solution integrated for 40 million years. Right: Amplification (40 times) of a small arc (Reproduced from Ferraz-Mello (1994b) with permission of the AAS)

2.8 The Hecuba Gap and the Zhongguo Group

The Hecuba gap is the one occurring in the 2:1 resonance with Jupiter (see Fig. 2.2), where the asteroids have periods close to 5.93 years (i.e., half Jupiter's period) (see Fig. 2.5). The name comes from the asteroid (46) Hecuba, discovered in 1857, by Karl Robert Luther, the first to be discovered close to this resonance (but outside it). In fact, the first object discovered inside this resonance was (3789) Zhongguo (or 1928 UF), in 1928 by Zhang Yuzhe. However, this asteroid was lost and remained so, until its rediscovery, with the ESO telescopes, in 1986. Meanwhile, the name given to it after its discovery, (1125) China, was attributed to another asteroid discovered in 1957 and incorrectly identified with the lost asteroid 1928 UF. It was then renamed as (3789) Zhongguo, the name of China in its official language, to preserve some memory of the events.[6] While Zhongguo was lost, the primacy of the gap remained with (1362) Griqua, discovered in 1935, by Cyril Jackson.

We can initially study the motions in this gap using the simplest model in which the asteroid is only influenced by Jupiter, which is moving in a fixed elliptical orbit, with the asteroid making 2 revolutions around the Sun while Jupiter makes one. The 3 most adequate angular coordinates to describe this 3-degree of freedom system while preserving the symmetries of the equations of motion are:

$$\begin{aligned} \psi &= \lambda - 2\lambda', \\ \theta &= \psi + \varpi, \\ Q &= \lambda - \lambda'. \end{aligned} \tag{2.6}$$

(see Sect. 2.5). The angle Q varies rapidly and should be eliminated using an averaging or a digital filtering method.

To construct the Surface of Section, we proceed as in the previous case, marking all intersections of the solution with the plane defined by $\theta = 0$ and representing the intersections using as polar coordinates the eccentricity of the orbit at the moment of intersection and the longitude of the perihelion (referred to the perihelion of Jupiter).

The Surfaces of Section shown in Fig. 2.22 were obtained using numerical integrations over 1 million years, followed by digital filtering of the high frequencies (see Sect. 2.7). These figures show that the regimes called low and medium eccentricities in the study of the Alinda gap are generally absent, appearing only in the last sections, which correspond to lower eccentricity values at the center of the resonance. In general, what we have is a chaotic region of low eccentricity ($e < 0.2$), confined, at medium eccentricities ($0.2 < e < 0.5$), by regular curves that surround it. The high eccentricity (or corotation) regime of the Alinda gap has, on the Hecuba

[6] We have chosen to retain the name Hecuba for the gap, although this asteroid is outside the resonance, because this name has been used extensively in the literature. In the case of the 3:1 resonance, we chose the name Alinda because the old name, Hestia, was not widely used and is known to few.

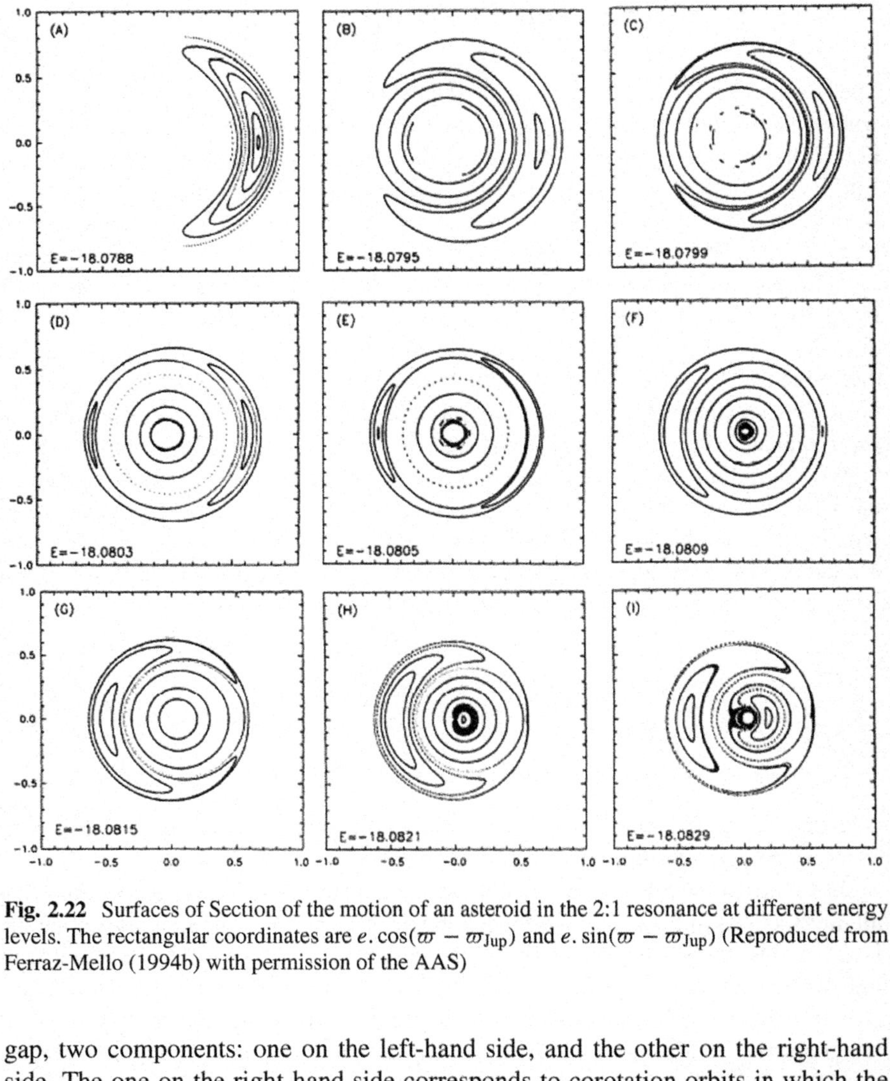

Fig. 2.22 Surfaces of Section of the motion of an asteroid in the 2:1 resonance at different energy levels. The rectangular coordinates are $e.\cos(\varpi - \varpi_{\mathrm{Jup}})$ and $e.\sin(\varpi - \varpi_{\mathrm{Jup}})$ (Reproduced from Ferraz-Mello (1994b) with permission of the AAS)

gap, two components: one on the left-hand side, and the other on the right-hand side. The one on the right-hand side corresponds to corotation orbits in which the asteroid perihelion oscillates around Jupiter's perihelion, while on the left-hand side, the asteroid perihelion oscillates around Jupiter's aphelion. The separation between these high-eccentricity motion regimes and the circulations below them does not exhibit signs of chaoticity. However, as shown in Sect. 2.10, their study with the Lyapunov characteristic exponents shows that the motions in this region have an important chaoticity (but which would require much more than 1 million years to manifest itself and be seen in the Surfaces of Section).

Let us comment on these figures. The first observation is that, in the inner parts, the eccentricities do not vary much along one solution. The closed curves seen have the appearance of slightly eccentric circles. The difference in the eccentricities on one side and the other of the figure, that is, when $\varpi - \varpi_{\mathrm{Jup}}$ is equal to 0° or 180°, is not large. This means that any asteroid whose initial conditions correspond to one of these curves never meets Mars. The mean distances of these asteroids and Mars to the Sun are 3.27 and 1.5 AU, respectively. Therefore, for an asteroid at this distance to reach the orbit of Mars, its eccentricity must be at least 0.6. Looking at the figures, we see that this possibility is remote. Any initial condition taken in the inner part of the Surfaces of Section results in an orbit that remains there.

Integrations starting from low eccentricities and over very long times, 80 million years, have been done on this model, and none of them led to eccentricities as large as the one needed to reach the orbit of Mars. The solutions exhibit quasi-periodicity aspects all the time.

In the model with Jupiter in a fixed orbit, there exists in the 2:1 resonance a chaotic region of low eccentricity confined by regular motions. In a model with two degrees of freedom, this confinement implies that it is impossible to leave this chaotic region. To get out of this region, a solution would have to cross the tori where the solutions are regular. It is impossible to come out of a tube without crossing its surface.

The inner chaotic region is a region with rich dynamics. In the points on the Surface of Section (Fig. 2.23), we see 4 empty holes near the outer edge and 3 near the inner edge. In the inner part of these holes, centers must exist that correspond to two stable periodic orbits: one near the outer edge, making four turns before closing (crossing the Surface of Section through the centers of the four holes in sequence), and another near the inner edge making three turns before closing. Around these centers, islands must exist formed by the intersections of regular motions lying on tori surrounding the stable periodic orbits. The chaotic region is formed by motions in the transition zone in which oscillations around these two periodic motions are

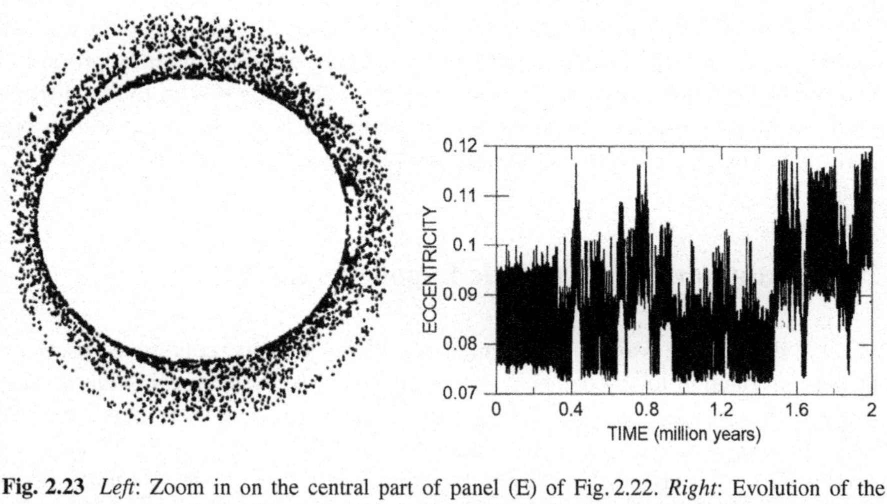

Fig. 2.23 *Left*: Zoom in on the central part of panel (E) of Fig. 2.22. *Right*: Evolution of the eccentricity in this solution (Reproduced from Ferraz-Mello (1994b) with permission of the AAS)

mixed. Using the results of the Poincaré-Birkhoff theory (Sect. 1.19), we can say that if there are four centers near the outer edge, there must also be four saddles. Likewise, near the inner edge there must be 3 saddles. From these 4+3 saddles, 28 asymptotic banches emanate, which entangle and give rise to a strongly chaotic region: all points in the figure belong to the same solution. To better understand what is going on, let us look at the same solution and see the graph showing the evolution of the eccentricity (recall that, in this Surface of Section, the eccentricity is given by the distance from the point to the center of the figure). We see that the eccentricity has epochs in which it is larger and epochs in which it is smaller, epochs that follow one another. If we had a lower level of chaoticity in this region, we could locate the 4+3 saddle points of this section and follow the asymptotic branches emanating from them. These branches would separate two regimes of motion: one of solutions with period 4 and the other of solutions with period 3. Chaos would be visible only in the region separating the two modes of oscillation, with solutions being able to move from one mode of oscillation to the other and mixing (as the intermittencies discussed in Sect. 1.18). But the chaoticity in this region is strong, and there is no such boundary: the two modes of motion mix completely. In summary, we have two modes of oscillation happening concurrently whose domains overlap, causing the strong chaoticity observed. To complete the analysis of this figure, we observe in panel (E) of Fig. 2.22 that, in addition to these periodic solutions, there is another periodic solution external to the discussed chaotic entanglement, crossing the Surface of Section at the centers of 5 islands.

The mechanisms responsible for the emptying of the 3:1 resonance are not repeated in the 2:1 resonance and do not allow us to explain the existence of the Hecuba gap. If we want to make progress in the study of first-order resonances like 2:1 and 3:2, we have to improve the model by incorporating other large planets (at least Saturn) and take into account the fact that the motions are not coplanar. It is just that when we do this, the number of degrees of freedom of the system increases. The previous system had a structure in which there was a high frequency, a medium frequency, and a low frequency, and by an averaging or digital filtering process, it was possible to eliminate the high frequency and reduce the system to two degrees of freedom. This will no longer be possible in models involving more than one perturbing planet, and new methods of chaos diagnosis need to be introduced.

2.9 Lyapunov Characteristic Exponents (LCE)

We will deepen the issue of exponential sensitivity to initial conditions discussed in the introduction to the definition of chaos and see how it can provide another method to diagnose chaoticity. As we have already observed more than once, we will be more concerned with the qualitative and operational aspects than with demonstrations. Exponential sensitivity indicators appear in the literature under several names: Lyapunov numbers, Lyapunov characteristic numbers, Lyapunov exponents, or simply LCE (Lyapunov characteristic exponents). The inverse of the largest LCE is usually called "Lyapunov time". These indicators are conceptually based on a series of theorems demonstrated by Russian mathematicians (Valery Oseledec, Yakov Pesin, etc.) and widely disseminated in the West by Giancarlo Benettin, Luigi Galgani and Jean-Marie Strelcyn (1976). We begin this section by presenting a method that is not the best, it even presents an often mediocre performance, but it makes use only of intuitive concepts.

Since chaos manifests itself as an asymptotically exponential divergence of two neighboring trajectories, to study it, we can construct solutions of the differential equations of motion starting at two nearby points, $\mathbf{x}$ and $\mathbf{x}'$ and follow them. When we are in a chaotic region, the tendency of these solutions is to move exponentially away from one another. The distance between the two trajectories in phase space tends to grow exponentially (see Fig. 1.3). But the exponential divergence itself conspires against such a simple method: if the two trajectories move apart exponentially, after a certain time, they will be so far away from one another that the distance between them no longer has any meaning. A simple example demonstrates this: Suppose that the trajectory in phase space is inside a sphere. Then, two paths on this sphere cannot move apart indefinitely because the maximum distance that two points can move apart is the diameter of the sphere. Therefore, if we follow two chaotic trajectories on a sphere, the distance between them can grow exponentially, but only until its value approaches the diameter of the sphere. If this fact is not taken into account, when we compare one trajectory with another, we will have meaningless results. This extreme example shows that there is a need to refine the idea, as discussed in Chap. 1 (Sect. 1.2).

We are trying to detect the chaos behavior along a trajectory; the neighboring trajectory has the sole purpose of exploring space in the vicinity of the investigated trajectory. When it moves too far away from the investigated trajectory, it ceases to do it.

Let us observe Fig. 1.4. There we have two solutions whose conditions differ by a tiny amount, and progressively they move away from one another. In the beginning, the two almost overlap, but gradually they move apart until after a certain point they are very far apart. After some time, the distance between the two solutions becomes too large and no longer serves to inform how these solutions diverge in the vicinity of one of them. Therefore, if the distance between the two solutions is being used to measure this divergence, one of them should be dropped and replaced by a new one closer to the solution under investigation. This is why Benettin et al., in their fundamental paper on the calculation of LCEs, indicate the need to normalize the distance to avoid the two calculated solutions being too far apart.

The recipe is simple: We calculate the trajectories starting at the points $\mathbf{x}(0)$ and $\mathbf{x}'(0)$. Let us call d_0 the distance (in phase space) between $\mathbf{x}(0)$ and $\mathbf{x}'(0)$. After a time τ, we stop the calculation. At this moment, the distance between the two points is d_1. Before restarting the calculation, we take the line joining the points $\mathbf{x}(\tau)$ and $\mathbf{x}'(\tau)$ and perform a renormalization that consists in taking as the starting point for the next step, instead of the point $\mathbf{x}'(\tau)$, one point $\mathbf{x}''(\tau)$ lying on the line joining $\mathbf{x}(\tau)$ and $\mathbf{x}'(\tau)$ and whose distance to the point $\mathbf{x}(\tau)$ is the initial distance d_0. (Fig. 2.24). We continue from this point and point $\mathbf{x}(\tau)$ until instant 2τ. At this

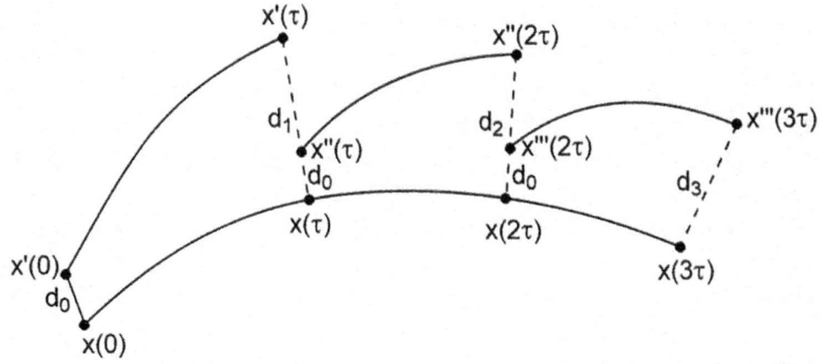

Fig. 2.24 Renormalization process

moment, the distance between the two paths is d_2. We stop the calculation and perform a new renormalization, taking as the starting point for the neighboring path between moments 2τ and 3τ, a point $\mathbf{x}'''(2\tau)$ which lies on the line connecting $\mathbf{x}(2\tau)$ to $\mathbf{x}''(2\tau)$ and whose distance to the point $\mathbf{x}(2\tau)$ is the initial distance d_0. And so we proceed until a time $n\tau$ sufficiently large to allow us to estimate the exponential divergence, which will be given by

$$\lambda(\tau, \mathbf{x}, d_0) = \frac{1}{n\tau} \sum_{i=1}^{n} \ln\left(\frac{d_i}{d_0}\right). \tag{2.7}$$

This result has some properties demonstrated by its proponents: (1) The result does not depend on τ. We can increase or decrease the renormalization time interval at will, as long as d_i never gets too large and the total interval does not change, without changing the result. If τ is too large, the two orbits will be too far apart, and the "neighboring" trajectory will no longer be in the neighborhood we want to study. On the other hand, if τ is taken too small, the quotient d_i/d_0 will always be close to 1 and the process will not evolve as desired, because the number of needed intervals will become too large and deteriorate the numerical precision of the process; (2) The result does not depend on $|d_0|$. Obviously there are criteria for the choice of d_0. The neighboring trajectory has to be really neighboring. d_0 cannot be chosen so large that the neighboring trajectory is not close to the investigated trajectory, nor so small that the two trajectories cannot be correctly distinguished one from another. If d_0 is too small, it will be confused with the errors of the numerical process, and we will be detecting the propagation of these errors and not the divergence of the two trajectories; (3) The limit

$$\Lambda_B(\mathbf{x}, d_0) = \lim_{n\to\infty} \lambda(\tau, \mathbf{x}, d_0) \tag{2.8}$$

is approximately equal to the LCEs (the error is of the order of the initial distance d_0).

2.10 Chaos and LCE

To understand the relationship between chaos and LCEs, let us do some math with a formula similar to the formula used to calculate λ, but without renormalizations:

$$\lambda(t) = (1/t)\ln d(t) \tag{2.9}$$

(we introduce $t = n\tau$ and change d_i/d_0 by $d(t)$). We will only study fictitious cases in which we attribute a certain behavior to the distance $d(t)$ between the paths. Let us consider the following examples:

$$\begin{aligned} &(A) \qquad d(t) = 1 + 10^{-5}t + \exp(10^{-5}t) \\ &(B) \qquad d(t) = 10^{-5}t + \exp(10^{-5}t) + \exp(10^{-3}t) \\ &(C) \qquad d(t) = 1 + 10^{-5}t + \exp(10^{-7.5}t) \\ &(D) \qquad d(t) = 2 + 10^{-6}t \end{aligned} \tag{2.10}$$

and calculate the quantity λ for 10^8 time units. The results are shown in Fig. 2.25. Before discussing this figure, let us calculate, for each case, the limit $\lim_{t\to\infty}\lambda(t)$.

They are trivial. In all cases, when $t \to \infty$, the quantities $\exp(kt)(k > 0)$ and their logarithm tend to ∞. Since to calculate λ we need to divide this quantity by t, which also tends to ∞, we need to perform an operation of the type ∞/∞. Such limits need some care. These are simple: all terms of type $\exp(kt)$ tend to ∞, and the result of the operation is dominated by the one that tends to ∞ faster. In the first 3

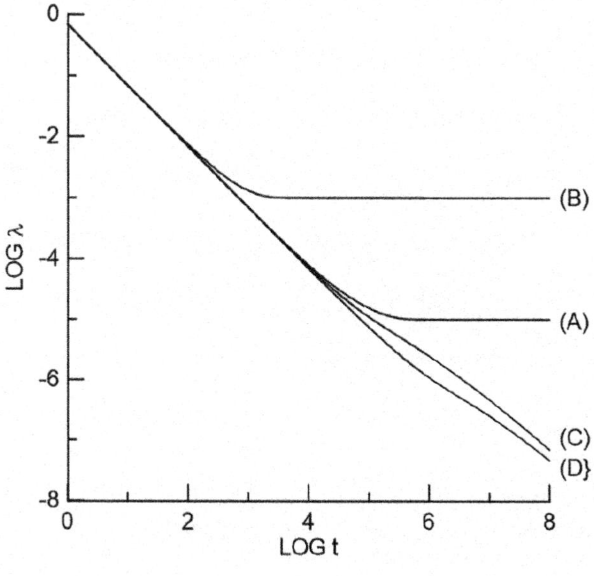

Fig. 2.25 Fictitious examples of the behavior of $\lambda(t)$

cases, the dominant term is the exponential with the largest coefficient k multiplying t; in case D, the term $10^{-6}t$ dominates. By calculating the logarithm and dividing by t we obtain, in the first 3 cases, the coefficient of the dominant exponential itself, and in case D, a function tending to 0. Thus λ tends to 10^{-5} in case A, to 10^{-3} (which is greater than 10^{-5}) in case B, to $10^{-7.5}$ in case C, and to 0 in the last case. Case D is a trivial exercise in calculating limits: applying the l'Hospital's rule to it, we obtain that the limit of λ when $t \to \infty$ is equal to the limit of $1/t$; this detail is important since it not only tells us that λ tends to zero but also how it tends to zero.

In the figure, what is shown is the (decimal) logarithm of λ, so we should expect that, in the four cases studied, the curves tend to -5, -3, -7.5, and $-\infty$, respectively. Let's start with the last one. In case D, the curve at the very beginning slopes downward and tends to a decreasing straight line that tends to form an angle of 45° with the horizontal axis, that is, $\log\lambda \sim -\log t$, according to the fact that in this case, at infinity, λ becomes inversely proportional to time. In cases A and B, the initial curve goes downward, but tends to become horizontal at the levels corresponding to the limits -5 and -3, respectively.

In case C, the behavior should be equal to these two, with the curve initially descending and becoming horizontal at level -7.5. However, the calculation for 10^8 time units was not enough for this level to be reached, and there is no indication that the curve will flatten out. In practice, a calculation of $\lambda(\tau, x, d_0)$ showing a descending straight line on the log-log diagram leads to the same dilemma. In the same way that we would not be able to distinguish, in the figure shown, if the curves were not marked, which one corresponds to case C and which one corresponds to case D. In a real case of calculation of $\lambda(\tau, x, d_0)$, we can never know if such a result indicates that the motion is either a regular motion or a chaotic motion with a limit value Λ below the minimum achieved in the calculated time interval.

Figure 2.26 shows some examples of functions $\lambda(t)$ (see Eq. 2.7) corresponding to several solutions of a system with two degrees of freedom that describes the

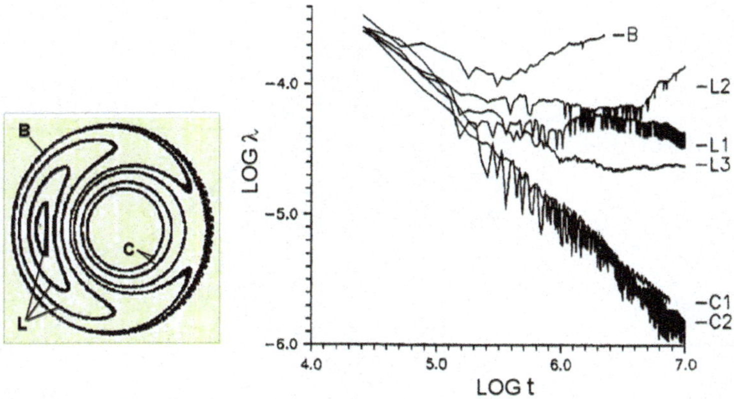

Fig. 2.26 Log-log diagram of the functions $\lambda(t)$ corresponding to several solutions of a real problem with two degrees of freedom whose Surface of Section is shown in the box at left. Time counted in years (see Sect. 2.14)

evolution of the orbit of an asteroid in a resonance calculated as indicated in Sect. 2.9. The Surface of Section of this system is shown next to the graphs and allows the various solutions studied to be identified. The coordinate axes of the Surface of Section are $e.\cos(\varpi - \varpi_{\rm Jup})$ and $e.\sin(\varpi - \varpi_{\rm Jup})$. In solutions C, the angle $\varpi - \varpi_{\rm Jup}$ circulates, and in solutions L (crescents), it oscillates around 180°; L_1 is the innermost one. Solution B is the outermost solution on the Surface of Section and shows intermittency of motions of types L and C. Due to this intermittency, solution B is the most chaotic with LCE greater than $10^{-4}\,\rm yr^{-1}$ (Lyapunov time less than 10,000 years). The solutions L are also chaotic but with LCEs smaller than $10^{-4}\,\rm yr^{-1}$. One can note that the solutions L appear blurred on the Surface of Section; one can expect that, in a longer integration, the various L-curves will mix. The solutions C do not show chaoticity. They have LCEs smaller than $\sim 10^{-6}\,\rm yr^{-1}$ and, in the time span of the experiment shown, behave as regular.

2.11 The Theory of LCEs. Variational Equations

We consider a point $\mathbf{x}$ on the phase space (of dimension $2N$) and another point $\mathbf{x}'$ in its neighborhood, obtained by giving to $\mathbf{x}$ an infinitesimal displacement $\mathbf{e}$. The variational equations of the given trajectory are obtained in a simple way. If the given differential system is written

$$\dot{\mathbf{x}} = \mathbf{X}(\mathbf{x}, t), \tag{2.11}$$

to obtain the variational equations of a trajectory $\mathbf{x}(t)$, it is enough to replace, in the given equations, $\mathbf{x}$ by $\mathbf{x} + \mathbf{e}$:

$$\dot{\mathbf{x}} + \dot{\mathbf{e}} = \mathbf{X}(\mathbf{x} + \mathbf{e}, t) \tag{2.12}$$

or, taking the linear approximation of the right-hand side (i.e., the first two terms of its Taylor series):

$$\dot{\mathbf{x}} + \dot{\mathbf{e}} = \mathbf{X}(\mathbf{x}, t) + \frac{\partial(\mathbf{X})}{\partial(\mathbf{x})}\mathbf{e} \tag{2.13}$$

where $\partial(\mathbf{X})/\partial(\mathbf{x})$ is a matrix (Jacobian). Eliminating the terms that correspond exactly to the given equation, we obtain the variational equation

$$\dot{\mathbf{e}} = \frac{\partial(\mathbf{X})}{\partial(\mathbf{x})}\mathbf{e}. \tag{2.14}$$

To calculate the LCE, we have to obtain the solutions of the given equations and of the variational equations simultaneously. The solution of the equations of motion gives the evolution of the vector $\mathbf{x}(t)$, and the solution of the variational equations gives the evolution of the vector $\mathbf{e}(t)$. Once this is done, we apply the Oseledec's formula:

$$\lambda(\mathbf{x}, \mathbf{e}, t) = \frac{1}{t} \ln ||\mathbf{e}(t)|| \tag{2.15}$$

and, as before,

$$\Lambda(\mathbf{x}, \mathbf{e}) = \lim_{t \to \infty} \lambda(\mathbf{x}, \mathbf{e}, t). \tag{2.16}$$

An Oseledec theorem guarantees that $\Lambda(\mathbf{x}, \mathbf{e})$ exists. This existence theorem is necessary since the calculated quantity tends to infinity in the numerator and in the denominator, and the limit of the quotient could do not exist. No renormalization is required when using Oseledec's formula because the variational system is linear in the variable $\mathbf{e}$ with coefficients that depend only on $\mathbf{x}(t)$. Therefore, a renormalization would do nothing to change the result.

This simplified presentation implicitly assumes that the phase space is Euclidean, which, in general, is not true. A rigorous presentation should consider that the phase space is not Euclidean, and that the displacement (variation) $\mathbf{e}$ of $\mathbf{x}$ must be defined in the Euclidean space tangent to the phase space at point $\mathbf{x}$ (Froeschlé 1984). Another important mathematical constraint to be remembered when one uses LCEs, is the following: the theorems allowing one to interpret positive LCEs as indicative of chaos are only valid if the hypersurfaces of the phase space formed by points of the same energy are compact. For instance, they cannot be infinite, and the discussed LCE properties do not apply if the solutions can go to infinity (in the case of mechanical problems, as the phase space encompasses positions and velocities, the properties do not apply to trajectories that can go to infinity or whose velocities can tend to infinity).

The result obtained with this definition is not very different from that obtained with the two-path formula of Benettin et al. As they have proved, the limit $\Lambda_B(x, d_0)$ reproduces the Oseledec LCEs to within one error of the order of the initial distance d_0. If d_0 is quite small, we can neglect this error. However, if the LCE is very close to zero, it does not allow to obtain its value nor discern whether it is positive, negative, or null. If, for example, we obtain that $\Lambda_B(x, d_0) = 10^{-8}$, if $d_0 \sim 10^{-8}$, we cannot interpret the result. Note that we are not talking of an error but of the fact that the formula of Benettin, Galgani, and Strelcyn does not exactly reproduce Oseledec's LCE definition formula. In such an extreme case, we need to use the variational equations and Oseledec's formula. There are also practical differences: the curve in the log-log diagram that gives the variation of the quantity λ appear smoother when the variational equations are used.

2.12 The Maximum Lyapunov Exponent (mLCE)

Let us now note that the results of the operations performed with the Oseledec formula are not unique. In fact, they depend on the initial displacement $\mathbf{e} = \mathbf{x}' - \mathbf{x}$. A second theorem of Oseledec states that $2N$ different and independent results can be obtained by properly choosing the initial displacements $\mathbf{e}(t_0)$. We can find $2N$ independent vectors $\mathbf{e}_1, \mathbf{e}_2, \mathbf{e}_3, \cdots, \mathbf{e}_{2N}$ (a basis) such that each of them leads to a different value of $\Lambda(\mathbf{x}, \mathbf{e})$. We have $2N$ different characteristic exponents: $\Lambda_1, \Lambda_2, \Lambda_3, \cdots, \Lambda_{2N}$.

Two important complementary results are: (1) At least one of the LCEs of the solutions of an autonomous system is equal to zero. The proof of this property is based on the following fact: if the system is autonomous, that is, if the "forces" are not depending on time, two solutions passing through the same position in the phase space at different moments of time may follow the same trajectory and the vector $\mathbf{e}(t)$ connecting simultaneous corresponding points in the two solutions cannot grow or decrease indefinitely; (2) The sum of the LCEs of a system satisfying the invariant property of Liouville (see Sect. 1.8) is equal to zero (Benettin et al. 1980). Indeed, if the volume must be preserved and if there are directions in which the system undergoes a strong expansion, then there must be other directions in which it undergoes a strong contraction so as to compensate them. The latter property is perhaps the most important characteristic differentiating conservative systems from dissipative systems. In dissipative systems, it can happen that all LCE are negative or zero. Attractors can exist, and the system can evolve by shrinking and tending toward an attractor. If there is only one attractor, all solutions in a large domain of initial conditions will tend to this attractor. In a conservative system, this can never happen.

In practice, it is very difficult to isolate exactly each of the vectors $\mathbf{e}_i$ leading to independent LCEs (see Sect.2.12.1). To explain this, let us consider a simple planar example (not autonomous since the two LCEs of a conservative system, which is autonomous and planar, must be equal to zero). Let $\mathbf{e}_1, \mathbf{e}_2$ be two vectors forming a base. Let us also suppose, to fix the reasoning, that the vector $\mathbf{e}_1$ tends to dilate and the vector $\mathbf{e}_2$ tends to contract (in the limit for $t \to \infty$). The LCE Λ_2 obtained from a vector oriented in the direction of $\mathbf{e}_2$ will be negative, and the other, Λ_1, obtained from a vector oriented in the direction of $\mathbf{e}_1$, will be positive. Then, integrating the differential equations from $\mathbf{x}$ and the variational equations from $\mathbf{e}_2$, we should obtain the negative LCE. But it happens that, in practice, it is not possible to know exactly the vector $\mathbf{e}_2$. No matter how good the approximation used for $\mathbf{e}_2$ the solution of the variational equation never starts exactly from a vector oriented in the direction of $\mathbf{e}_2$. Even though we have good information about $\mathbf{e}_2$, its practical implementation is an initial displacement $\mathbf{e}$ which is a composition of $\mathbf{e}_1$ and $\mathbf{e}_2$ (see Fig. 2.27). The projection of $\mathbf{e}(t)$ in the direction of $\mathbf{e}_2(t)$ tends to contract, and the projection in the direction of $\mathbf{e}_1(t)$ tends to dilate. After a large number of iterations, the resultant vector $\mathbf{e}(t)$ will tend to align with $\mathbf{e}_1(t)$. As the solution is computed, the projection onto $\mathbf{e}_2(t)$ will decrease, and the alignment with $\mathbf{e}_1(t)$ will become dominant. Since

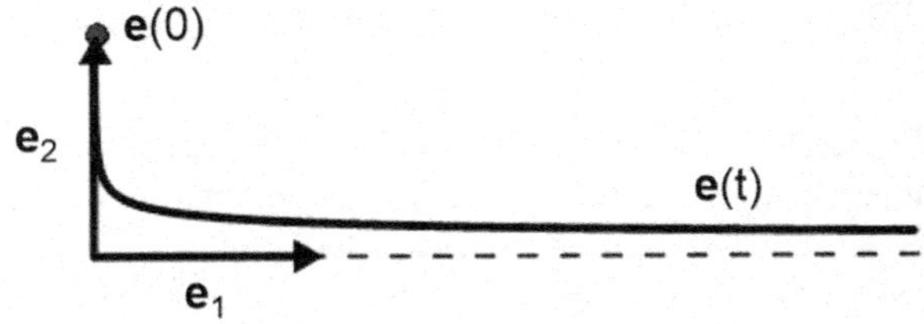

Fig. 2.27 Dominance of the largest LCE

the LCEs are obtained by a process where $t \to \infty$, any initial vector that is not absolutely coincident with one of the basis vectors will be pulled in the direction of the greatest divergence.

In conservative systems, we always have directions in which the system lengthens and others in which it contracts. The direction in which the system lengthens dominates. Therefore, in practice, the numerical integration of the variational equations always gives the largest LCE as a result.

In principle, the result of the operation proposed by Oseledec depends on the position where the studied trajectory begins. In a system where there are KAM tori, a solution on a torus should have the largest LCE equal to zero, while a solution outside of them, where a homoclinic entanglement is acting, should have at least one positive LCE. Likewise, the result will not be the same in one zone of regular solutions and another of chaotic solutions; or in two separate zones of chaotic solutions. However, within the same chaotic region, the result is independent of $\mathbf{x}_0$. For instance, in the Wisdom regime of the first figure of the 3:1 resonance (Fig. 2.13) all points inside the oval are intersections obtained from the same initial position. For any starting point within this region, the largest LCE found should be the same. In this case, something close to $0.0003\,\text{yr}^{-1}$ (Wisdom 1983).

In general, in a chaotic region, the result obtained does not depend on the initial condition. However, in cases of slower chaoticity, the fact that the LCEs are defined from a limit operation to $t \to \infty$ means that if the time interval used to calculate the LCEs is not long enough, the results obtained may be different. This is an important limitation that makes the LCE calculation unreliable in problems with very slow chaoticity. The results may be considered as indicators of local character (but they are no longer the LCE). Trajectories computed for short times can lead to results very different from the actual LCE.

2.12.1 *Calculation of the Other LCEs*

If the largest of the LCE draws the result, whatever the initial value of **e**, how to calculate the other LCEs? We can think, for example, that taking the initial vector exactly perpendicular to the initial vector corresponding to the largest LCE, the result should be the second largest LCE. But in practice, it is impossible to obtain such orthogonality exactly. We then proceed as follows (Benettin et al. 1980):

The variational equations are integrated considering initial unitary displacements mutually orthogonal, for example, $\mathbf{e}_1, \mathbf{e}_2, \mathbf{e}_3, \ldots$, and we compute their evolution. As it has been discussed, they will all be driven in the direction of the largest divergence of the system. None of them will be kept strictly perpendicular to that direction. To get around this fact, a reorthogonalization and renormalization process is necessary. We integrate the equations of motion for each of the initial conditions $\mathbf{e}_1, \mathbf{e}_2, \mathbf{e}_3, \ldots$, for a certain time τ; because of the dominance of one of the components, this set of initially orthogonal vectors will lose this property. It is then reorthogonalized and renormalized. This reorthonormalization is done as in a Gram-Schmidt process, starting from the largest vector to the smallest. The largest vector is just renormalized (its size is reset to 1). The second largest vector is renormalized and rotated through the plane formed by it and the largest vector until they are perpendicular. The third vector is renormalized and rotated in the same way until it is perpendicular to the first two vectors (this is done by subtracting from it its projections on the first two vectors). And so on. Once all vectors are reorthonormalized, we integrate the equations for another interval τ, and at the end of it proceed at a new reorthonormalization. And so forth. The successive reorthonormalizations allow us to obtain information about all the LCE. This is extremely expensive from a computational point of view.

2.13 Exponential Divergence and Information Loss

Suppose that the solutions of a system diverge with an exponent λ. Any initial displacement d_0 will grow as shown in Fig. 1.3. Let us remove the fluctuations and consider only the exponential. In this case, the Lyapunov time, which is the inverse of the mLCE, represents the time required for the initial displacement to increase by e times ($e = 2.71828\cdots$). We can change the logarithm basis used in the definition of λ from the basis e to the basis 2, which is the usual basis in information theory. To obtain the time in which the initial displacement is multiplied by 2, it is sufficient to multiply the Lyapunov time by $\ln 2 = 0.693\,147\cdots \sim 0.7$. That is, the displacement vector doubles in size every 0.7 Lyapunov times. If we write the initial value using a binary representation, the indetermination of the initial value is a sphere with a radius equal to the last known bit. In 0.7 Lyapunov times, this inaccuracy doubles, that is, one bit of information is lost. At 2 times 0.7 Lyapunov times, another bit is lost, and so on. This poses, from a practical point of view, a limitation from which we cannot escape.

If we repeat the above reasoning using the basis 10 instead of 2, we find the time at which the initial displacement is multiplied by 10 by multiplying the Lyapunov time by $\ln 10 = 2.302\,585\cdots \sim 2.3$ In other words, the displacement vector decuples every 2.3 Lyapunov times. This means that the inaccuracy of the solution grows tenfold every 2.3 Lyapunov times and that a decimal digit is lost every 2.3 Lyapunov times. As an example, the dynamical study of Hyperion's rotation gives a Lyapunov time of 40 days (or mLCE $= 0.025\,\mathrm{d}^{-1}$) (Shevchenko and Kouprianov

2002; Shevchenko 2020). Thus, in that problem, inaccuracies are decupled every 3 months (see Sect. 1.4).

The numerical simulations of the evolution of dynamical systems have insurmountable limitations that cannot be overcome by improved algorithms. They are inherent to chaotic systems and independent of the algorithm used. It is never possible to use all the digits in a significant way. The last ones will always lose importance. There are rules of thumb like this: If the computer works with numbers made up of 15 significant digits, there is no point in asking for results with a precision better than 11 or 12 digits. But even if all digits could be used and a perfect integrator was available, in the case of chaotic solutions there exists a limitation coming from the exponential growth of the initial inaccuracies.

2.14 Application to Resonant Asteroids

Let us discuss the problems of the motion of an asteroid in resonance with Jupiter and of Kirkwood gaps in the light of Lyapunov exponents. Let us first consider the solutions obtained for the 2:1 resonance using digital filtering to reduce the system to 2 degrees of freedom, in the case shown in panel (E) of Fig. 2.22. We use the letters $C_1, C_2, C_3, \cdots, C_7$ to designate the solutions of the figure in which the perihelion is circulating the origin of coordinates, from inside to outside; C_1 is the chaotic solution confined in the central part also shown in Fig. 2.23. Let us denote with LL and LR the two solutions in which the perihelion is oscillating on the left and right sides of the figure. Let us compute these solutions and those of neighboring (renormalized) trajectories for 10 million years and, with the help of the Benettin, Galgani, and Strelcyn formula, calculate the λ-function (see Eq. 2.7). Let us start with the solution C_1. The corresponding curve on the diagram $\log \lambda$ *vs.* $\log t$ is the one shown at the top of the first panel of Fig. 2.28. Looking at its trend, we see that the mLCE corresponding to this curve, that is, the limit of $\lambda(t)$ for increasing t, is $\sim 10^{-3.5}\,\mathrm{yr}^{-1}$ (Lyapunov time $\sim$ 3000 years). In contrast, for solutions C_2 and C_3, the curves $\lambda(t)$ decrease all the time; it seems that we can guess an inflection starting at their ends in the log-log diagram, but the computation time was not enough to define its nature. To do it, we would need even longer integrations. The limit of $\lambda(t)$ for increasing t in these cases is smaller than $10^{-5.5}\,\mathrm{yr}^{-1}$ indicating Lyapunov times larger than 300,000 years. The curves corresponding to solutions with oscillating perihelions (LL and LR) show inflection and horizontalization indicating mLCE between 10^{-6} and $10^{-5}\,\mathrm{yr}^{-1}$, that is, Lyapunov times of 10^5–10^6 years.

With the LCE, we can more easily tackle problems with several degrees of freedom, for which it is not possible to obtain Surfaces of Sections represented in a plane. We can, therefore, redo the same calculations above for a more complete model in which not only Jupiter but also Saturn is present and in gravitational interaction with the other bodies considered. We can also leave aside the planar models and consider reality in its three dimensions. In this new model, the number of degrees of freedom is greatly increased. Surfaces of Section become useless, but

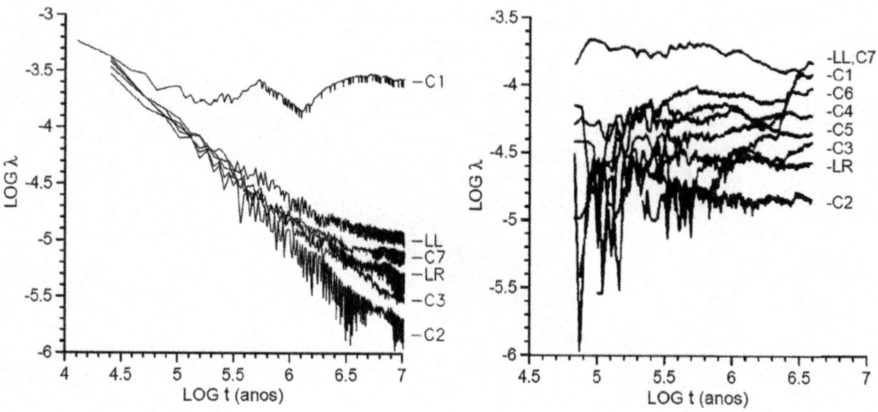

Fig. 2.28 Log-log diagram of the functions $\lambda(t)$ corresponding to several solutions for the motion of an asteroid in the 2:1 resonance corresponding to the Surface of Section (E) of Fig. 2.22. *Left*: 2D model with Jupiter in a fixed orbit. *Right*: 3D model with Jupiter and Saturn

the LCE and Lyapunov times can be calculated. The second panel of Fig. 2.28 shows the curves $\lambda(t)$ obtained from trajectories computed for 5 million years with the 3D model including Jupiter and Saturn. These curves show the strong chaoticity of all solutions. One of them has a very curious behavior, with a sudden rise at the end. Their limits for increasing t indicate mLCEs between 10^{-5} and 10^{-4} yr^{-1}, that is, Lyapunov times of 10^4–10^5 years.

2.15 Events. Sudden Orbital Transitions

An extensive program of numerical simulations of the motion of asteroids under the perturbation of one or more planets was conducted at Harvard and led by astronomers Myron Lecar and Fred Franklin. Analyzing the results of these simulations, they observed that the sudden orbital transitions that lead an asteroid to cross a planetary orbit occur in a time proportional to the Lyapunov time raised to a certain power (Lecar et al. 1992a,b). The formula given by them for objects moving between Mars and Saturn is, limiting the digits to the significant ones,

$$T_{\text{ev}} = 4T_{\text{Lyap}}^{1.8\pm0.1} \tag{2.17}$$

(times measured in years). The "event" or "sudden orbital transition" referred to in the formula may be a close encounter with a planet, the crossing of a planetary orbit, or the ejection of the system after a near collision. Soon after, other authors independently arrived at analogous results with powers varying, depending on the experiment, between 1.4 (Holman and Wisdom 1993) and 1.9 (Levison and Duncan

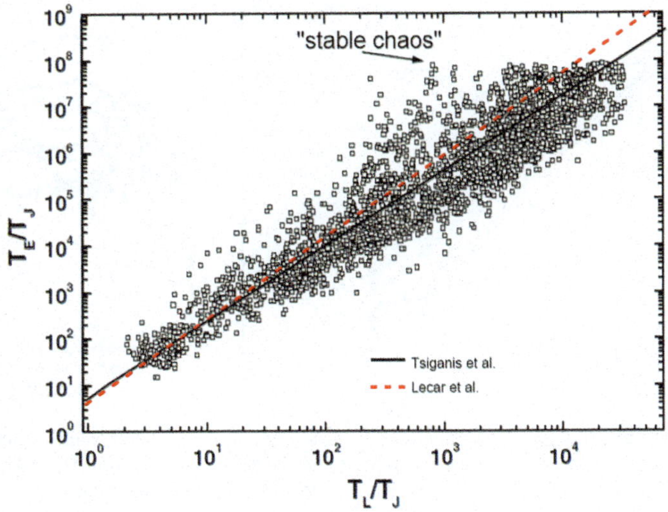

Fig. 2.29 Log-log diagram of the escape times (T_E) as a function of Lyapunov times (T_L) resulting from the integration of a large number of real and fictitious Trojan asteroids. The black line is the least squares fit of a power law to these points. The red line is the superposition to this figure of Lecar et al. power law. The arrow indicates a set of sample points with $T_L \sim 10,000$ years that lie well above the power law suggesting that these are cases of stable chaos. T_J is the orbital period of Jupiter (Adapted from Tsiganis et al. (2005). Copyright 1996 by Springer)

1993). But they were less incisive about the fatality of this empirical formula, emphasizing its mere statistical validity.

A more recent result involving a large number of objects was obtained by Kleomenis Tsiganis, Harry Varvoglis and Rudolf Dvorak (2005) from a study of a few thousand Trojan asteroids, real and fictitious, distributed on a grid enveloping the set of real objects (see Sect. 3.9). Fitting a power law to these points gives a result very close to Lecar et al.'s result:

$$T_{\rm esc} = (0.75 \pm 0.08) T_{\rm Lyap}^{1.62 \pm 0.03}. \tag{2.18}$$

The difference between these two results can be seen in Fig. 2.29, where the power laws found in the cited paper and by Lecar et al. are represented by a solid black line and a dashed red line, respectively. Figure 2.29 also shows the scattering of the obtained escape times increasing for higher Lyapunov times. This may be due to the increase in the indetermination of the Lyapunov times as they become larger (as seen in Figs. 2.28 and 2.36). Even though this study considered only solutions with good measures of $T_{\rm Lyap}$ to avoid adding a spurious scattering to the data.

The empirical formula $T_{\rm ev} = aT_{\rm Lyap}^b$ was contested by several groups. There is one fundamental criticism of the determinism with which Lecar et al.'s formula was first stated. The criticism is the following. Let us observe the Surface of Section of the 2:1 resonance. In their interior part, strong stochasticity occurs, and the calculated Lyapunov time is of the order of 5000 years (see curves C_1 in Fig. 2.28). The time for an event, by the given formula, is less than 10 million years. However, due to the robust structure of regular motions around this chaotic region in the model where only the perturbation due to Jupiter is considered (see Fig. 2.22) an object will not escape from it even in 100 million years. It does not matter what the Lyapunov time is in the inner chaotic region because it is confined, and a trajectory cannot escape from this region in a predictable time. This is a clear demonstration that a deterministic law relating Lyapunov times and sudden transition times cannot exist. Another conceptual restriction on the existence of such a formula is the essence of chaoticity after which two orbits that begin at neighboring points can have completely different developments. See the examples shown in Fig. 1.18 and the discussions concerning them.

In favor of Lecar et al.'s formula, it can be argued that the prohibitive analyses that we have just made were based on very simplified models. The motions of the real planets are not mere solutions of a planar restricted problem involving three bodies but occur in a real Solar System that has four large planets: Jupiter, Saturn, Uranus, and Neptune. In fact, when other planets are included in the model, the apparent regular solutions C_2 and C_3, of the restricted model, become strongly chaotic and no longer confine the solutions C_1. In a full model, orbits initiated in the inner chaotic region of the 2:1 resonance may escape at times much shorter than the age of the Solar System, via huge increases in eccentricity, as first shown by Hans Scholl (see Fig. 2.30). The complexity of systems is perhaps the reason why experiments done by various authors with different systems representing real problems in the Solar System have shown similar results with slightly different values of the exponent.

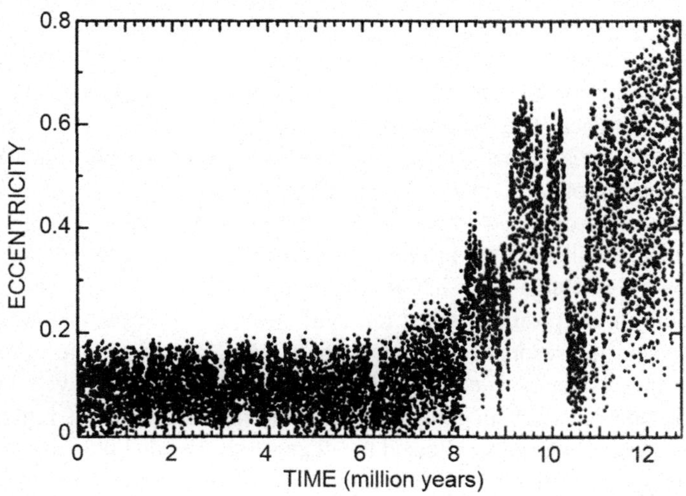

Fig. 2.30 Evolution of the eccentricity of a fictitious asteroid in the resonance 2:1 (cf. H. Scholl) (Reprinted from Ferraz-Mello (1994a) with permission of IAU)

The attempt to obtain information about escape times without having to make the numerical simulation of the system over all the time it takes to escape is an ambitious goal to pursue. With the formula of Lecar et al., one does it with an anticipation of a few orders of magnitude. In the case of the asteroidal resonances, the Lyapunov times used in the discussion were calculated over 10^7 years, which is less than one hundredth of the age of the Solar System. Some experiments done using Shannon entropy show the existence of a correlation between the escape times and the inverse of the diffusion coefficient for times up to 4 orders of magnitude greater than the time used to calculate the diffusion (Beaugé and Cincotta 2019; Cincotta et al. 2021).

2.15.1 Stable Chaos

The initial objection to the formula of Lecar and co-workers was based on an example: the asteroid (522) Helga. Helga is an important asteroid, discovered by Max Wolf in 1904, whose diameter is estimated to be about 100 km, and whose importance is because it is one of the few large objects found beyond the edge of the main belt of asteroids at a distance from the Sun of 3.6 AU. The Lyapunov time of the orbit of the asteroid Helga is very low, about 7000 years (Milani and Nobili 1993). Nevertheless, as the opponents of Lecar and co-workers' formula said, it is there despite the 4.57 billion years of the Solar System's age. And it also remains more or less quiet there in simulations of its motion for the next 1–2 Gyr. The two arguments are important but somewhat misleading: the first because the fact that it is there does not mean that it has always been there; the second because the claimed simulations (in 1993, when the discussion took place) were not long enough to reach the time $T_{\rm ev}$.

But the main difficulty with these arguments was to rely only on the simulation of Helga's motion. In fact, the Lyapunov times of some calculated orbits close to it were much higher. The asteroid Helga is near the 12:7 resonance, and in this resonance, there is an inner chaotic region, with Lyapunov times on the order of 7000 years, but around it, there are much less chaotic structures, with Lyapunov times on the order of 50,000 years. Helga may be a remnant of a larger group of asteroids that initially existed in this region (Murison et al. 1994). The study of a group of clones of Helga with similar ellipses just rotated with respect to the actual one and with initial positions taken randomly over these ellipses shows that most of them escape the resonance in less than 10 Myr, but about one third of them are yet on the resonance even in 50 Myr. The escape times follow a log-normal distribution so that, even in longer times, some specimens must remain on the resonance. These studies also allowed the clones to be divided into an escaping and a surviving set according to the initial position of their perihelions. Helga would be more or less in the frontier of these two sets (Tsiganis et al. 2000, 2002). The structures observed in the vicinity of Helga seem to persist even when more complex models are adopted, allowing the asteroid to remain in the 12:7 resonance even in simulations approaching one billion years

An aggravating factor in this discussion is the need of long simulations to prove or disprove the thesis that the Lecar et al. formula gives good results in the study of the real system. They must extend for 1000 Lyapunov times or more. What is the significance of simulations for such a long time? Would it be that after a few hundred Lyapunov times, the trajectory obtained by the computer still contains some information about the real solution? Or is it already mere fiction? We are in the realm of possible artifacts.

A better characterization of the phenomenon known as stable chaos appears in Fig. 2.29. We observe that, in this figure, the lower edge is well defined, while the upper edge clearly shows a greater spreading of the boundaries when $T_{\text{Lyap}} \sim 10^4$ yrs (i.e., in the notations of the figure, $T_L/T_J \sim 10^3$). This increase was credited by its authors to the occurrence of stable chaos.

2.16 Fast Lyapunov Indicators (FLI)

In general, the Lyapunov characteristic exponents demand long numerical integrations of the differential equations of motion and do not allow the construction of dynamical maps on grids that are both dense and extensive. The search for more expedite alternatives has been intense. One example of a successful alternative is MEGNO, a variant of the Lyapunov characteristic exponents widely used in the study of exoplanet systems (see Sect. 3.12). While a premature truncation of the Oseledec formula (Eq. 2.15) in the calculation of the Lyapunov exponents can lead to gross errors (see Sect. 2.10), the faster convergence of the variant used in the MEGNO calculation allows the calculation to be done over shorter time spans without the same risk of introducing large errors.

In many cases, chaos researchers have renounced the study of diagnostic indicators of chaos and limited themselves to local indicators of divergence capable of distinguishing different regimes of motion in highly chaotic systems. These indicators can reveal the presence of global fast transport structures formed by the entanglement of stable and unstable spatial manifolds, which include trajectories that can cover long distances in short times.

Since their formulation by Claude Froeschlé, Elena Lega, and Robert Gonczi (1997), various definitions have been used for the so-called Fast Lyapunov Indicators. In all cases, the basic function is the variation of the tangent vector $\mathbf{e}(t)$ obtained by integrating the variational equations of the studied system (see Sect. 2.11). In the first studies, the FLI was defined as the time required for some functions of $||\mathbf{e}(t)||$ to reach a previously set limit. Since $||\mathbf{e}(t)||$ is a fast-growing function, the chosen function rather considered its inverse, and the fixed limit was a lower limit. An example is the function $\Psi = 1/\sup||e_j(t)||^n$ used in early FLI applications in the study of asteroidal motions (Froeschlé et al. 1997). Around 2000, again for the sake of having faster calculations, the FLI was defined as the supremum of $||\mathbf{e}(t)||$, or of its logarithm, $\log ||\mathbf{e}(t)||$, in a pre-fixed time interval (Froeschlé et al. 2000). This is a reduction of Oseledec's formula (without the division by t), and its value tends to ∞ when time grows. It is only valid for short times and, unlike the LCE, the FLI thus obtained can only be used comparatively within the same dynamical map.

An important detail to mention is the dependence on the initial displacement $\mathbf{e}_0 = \mathbf{e}(t_0)$: different values are obtained if different initial displacements are used. This dependence also occurs in the calculation of the Lyapunov characteristic exponents, but it is much less important because of the long time spans used, and the progressive alignment of the solution of the variational equations with that corresponding to the maximum LCE (see Sect. 2.12). In the case of computations over very short times, as in the example considered in this section, the time span of the calculations is not sufficient for this alignment to occur, and the choice of $\mathbf{e}_0$ may require special attention. Applications were reported where the quality of the solutions was improved by considering the $\mathbf{e}(t)$ component perpendicular to the flow (Fouchard et al. 2002).

Figure 2.31 shows a dynamical map constructed by Nataša Todorović et al. (2022) with the FLI defined by $\sup_{t \le T} |\mathbf{e}(t)|$, on a domain of the Solar System from the edge of the main asteroidal belt to far beyond the orbit of Uranus. For that sake, the FLIs were computed for more than 2 million initial conditions taken on a grid with uniformly distributed semi-major axes and eccentricities. The initial mean longitude was taken 60° ahead of Jupiter, while the other orbital elements were those of Jupiter at a fixed date. The 7 planets considered in the calculations (from Venus to Neptune) were initialized at their actual positions on the same date. The results obtained after a numerical integration over 100 years for each initial condition were plotted using the color codes indicated on the side of the map.

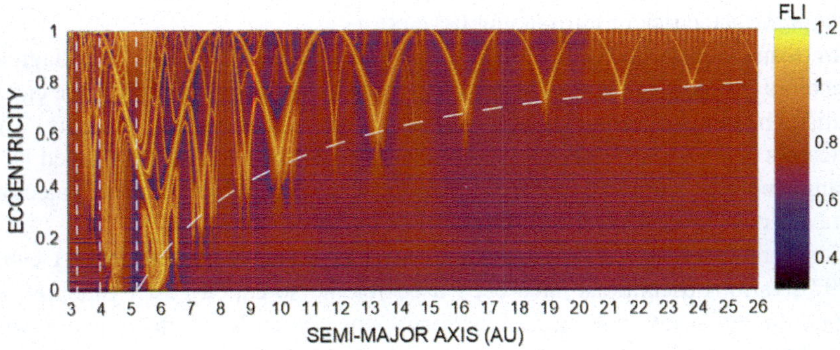

Fig. 2.31 Dynamical map of the Solar System showing the FLI values for more than 2 million initial conditions and revealing structures along which transport can occur at times of the order of orbital periods. The variable plotted on the graph is the FLI obtained from a 100-year integration of the variational equations. The lower dashed line indicates the initial conditions where the perihelion of the corresponding orbit lies close to Jupiter's orbit. The three vertical lines on the left show the loci of the asteroidal resonances 2:1, 3:2 and 1:1 (Trojans) © Nataša Todorović (Authorized reproduction)

The most notable formation is the chaotic V-shaped structure formed by the points corresponding to faster orbital divergence (light orange) visible near 5.6 AU and its connection to a series of arches at greater distances whose vertices follow the dashed line corresponding to orbits whose perihelions lie near the orbit of Jupiter. The arches lie above this line and correspond to orbits that reach Jupiter's orbit and may undergo major changes as they pass in the vicinity of the planet. Similar structures associated with the outer planets are also visible, although fainter. The niche where the Hildas are located (resonance 3:2) and the large island of stability in the vicinity of the Lagrangian point L_4, where the Trojan asteroids accumulate ahead of Jupiter, characterized by low divergence (blue), are also visible.

This intricate structure of instabilities acting on time scales of the order of the orbital periods of the planets appears only when short integration times (100 years) are used. Each region is, in fact, immersed in a chaotic ocean where these structures blend with others and become not visible when longer integration times are used. The spatial manifolds revealed by these FLI-maps are of the same nature as those used in the design of new spaceways and orbital configurations for spacecrafts destined for outer space exploration missions (Perozzi and Ferraz-Mello 2010).

Studies of fundamental dynamical models have revealed that, computed on short time intervals, finite-time chaos indicators allow to detect resonances, represent the phase portraits of complex dynamics, compute center-stable-unstable manifolds as well as other dynamical structures (Guzzo and Lega 2023).

2.17 The Hecuba Gap Asteroids

Let us see next how the facts considered so far show up in real asteroids. Figure 2.32 highlights the three groups of asteroids found on the Hecuba gap (Roig et al. 2002). First are (3789) Zhongguo and a large group of asteroids right in the middle of the resonance, in stable orbits. At high eccentricities, we have the Griquas. They are asteroids in marginally stable orbits that can remain in the 2:1 resonance for many millions of years (Roig et al. 2002; Brož et al. 2005). Furthermore, at higher eccentricities, the third group appearing in Fig. 2.32 includes asteroids in rather unstable cometary orbits (Roig et al. 2002; Brož et al. 2005) They have the Tisserand invariant $\mathcal{T} < 3$ and do not remain in the 2:1 resonance for more than 10^2–10^3 years.

The first asteroids discovered were continuously observed from their discovery, confirming, by observation, their librations around the central position of the resonance. The points Zh, Z_0 and Gr, G_0 represent the semi-major axes and eccentricities of (3789) Zhongguo and (1362) Griqua, currently and shortly after their discoveries (Berliner Astronomisches Jahrbuch 1932, 1937; Ferraz-Mello 1989), respectively. Besides these, we highlight in the figure the asteroid (108) Hecuba because it is the asteroid that gives its name to the gap.

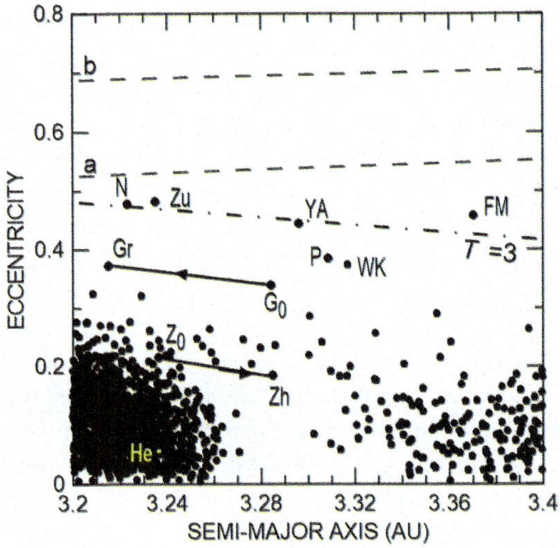

Fig. 2.32 Distribution of semi-major axis and eccentricity of the asteroids in the 2:1 resonance neighborhood with absolute magnitude up to 14. Zh, Z_0, and Gr, G_0 indicate the asteroids (3789) Zhongguo and (1362) Griqua today and at the time of their discovery one century ago. The lines connecting these points show the displacement of these asteroids due to their libration since their discovery. Also indicated are the high-eccentricity resonant asteroids (1921) Pala, (1922) Zulu, (3688) Navajo, (5201) Ferraz-Mello, (83943) 2001 WK14, (143243) 2002 YA26, and the asteroid (108) Hecuba. Lines a and b indicate the lower limits from which the asteroid's orbits cross the orbits of Mars and Earth, respectively. The other line corresponds to the position of the asteroids whose Tisserand invariant is $\mathcal{T} = 3$ ($\mathcal{T} > 3$ below the line)

To conclude, we can note that we did not include in this section a figure like Fig. 2.16 related to the Alinda gap. The reason is simple: the asteroids of the 2:1 gap are about 1.5 times farther from the Earth than those of the 3:1 gap. In addition, the asteroids on the Hecuba gap do not generally excursion through very high eccentricities and do not come very close to the Earth. For these reasons only a few faint asteroids, with absolute magnitude $H > 18$, have been discovered in this gap to date.

2.17.1 Zhongguo Group

The presence of a group of stable asteroids in the 2:1 resonance means that the Hecuba gap is not as empty as the Alinda gap (Roig et al. 2002). The most complete dynamical models show the existence of a zone of smaller diffusion with eccentricities in the interval $0.2 < e < 0.3$ and Lyapunov times of 10^5–10^6 years (Ferraz-Mello 1994b), which is populated by asteroids that can remain in the 2:1 resonance for times on the order of the age of the Solar System. The location of this zone depends on the perturbations of Saturn, especially those associated with the frequency of the great inequality of Jupiter's motion (Ferraz-Mello et al. 1998a). However, the diameters of these asteroids do not exceed 10–20 km, in contrast to the neighboring nonresonant asteroids which, in many cases, exceed 50 km and even 100 km in diameter. This is an indication that these asteroids should not be primordial. Moreover, the distribution of the diameters of these asteroids resembles that of asteroid families formed by fragments, which leads to believe that these small asteroids may have been injected into the resonance by the fragmentation of a large asteroid in their vicinity. Given the proximity of the (24) Themis family and the 2:1 resonance, Michèle Moons and co-workers (Moons et al. 1998) proposed that the same event that created that family injected a large number of fragments into the resonance. The Themis family is very old, billions of years old (Carruba et al. 2016), and the region is stable enough to preserve asteroids injected in ancient times until the present epoch.

2.17.2 The Griquas

At high eccentricities, we have the Griquas moving in marginally stable orbits with Lyapunov times of 10^4–10^5 years (Ferraz-Mello 1994b). The residence times of these asteroids found oscillating around the center of the resonance can be very different from each other and, strictly speaking, they form two subgroups. The group to which (1362) Griqua and (3688) Navajo belong has quite regular orbits and these asteroids can remain in resonance for hundreds of millions of years (Roig et al. 2002; Brož et al. 2005). Those of the other group, to which (1921) Pala, (1922) Zulu, and (83943) 2001 WK14 belong, remain in the resonance for much shorter times: 10–20 million years. In both cases, the residence times in the resonance are much shorter than the age of the Solar System, and these asteroids cannot be primordial. They are assumed to have penetrated the 2:1 resonance from its edges, driven by the Yarkovsky effect and, once in the resonance, to have migrated along the resonance edges to large eccentricities (Moons et al. 1998). The injection of small asteroid particles from collisions within the Themis family has also been invoked; the Themis family is very large and collisions may have occurred in recent times involving its members. An example is the (656) Beagle family, which formed about 10 million years ago (Carruba 2019) and may have injected new objects into the 2:1 resonance.

2.17.3 Resonant Asteroids in Cometary Orbits

The remarkable fact in this group is the value of the Tisserand invariant, smaller than 3, which is a characteristic of cometary orbits (usually for asteroids we have $\mathcal{T} > 3$). This rule is not strict, and the value of $\mathcal{T}$ is not sufficient to identify the nature of an asteroid, because there are many asteroids with $\mathcal{T} < 3$, and there are also comets with $\mathcal{T} > 3$. However, in the case of (5201) Ferraz-Mello, an asteroid discovered by Ted Bowell in 1983, observations made with SOAR (Southern Astrophysical Research Telescope) showed that the colors of this object are not normal for an asteroid. Rather, they are comparable to those of outer Solar System bodies like transneptunian objects and centaurs. On the other hand, similar colors are observed in cometary nuclei in the presence of a weak dust coma. These facts contribute to the supposition that, in fact, (5201) Ferraz-Mello is an extinct or dormant comet (Carvano et al. 2008).

The asteroids (1922) Zulu and (3688) Navajo also have Tisserand invariant values smaller than 3 and, therefore, cometary orbits, but simulations of their motion have shown that their orbits are nearly periodic and slow evolving, which rather places them among the Griquas.

Figure 2.33 shows how, in the 2:1 resonance region, the population of asteroids in cometary orbit mixes with the comets of the Jupiter family. The resonance ensures the robustness of this group by preventing the objects of coming too close to Jupiter when they are in the aphelion of their orbits (see Fig. 2.5).

Another fact shown in this figure is the motion predicted for the asteroid (5201) Ferraz-Mello in the next 500 years. We see that the semi-major axis and the eccentricity of the asteroid oscillate synchronously between two boundaries located at the edges of the resonance. This is the typical motion of asteroids located within a resonance and is what astronomers call a libration.

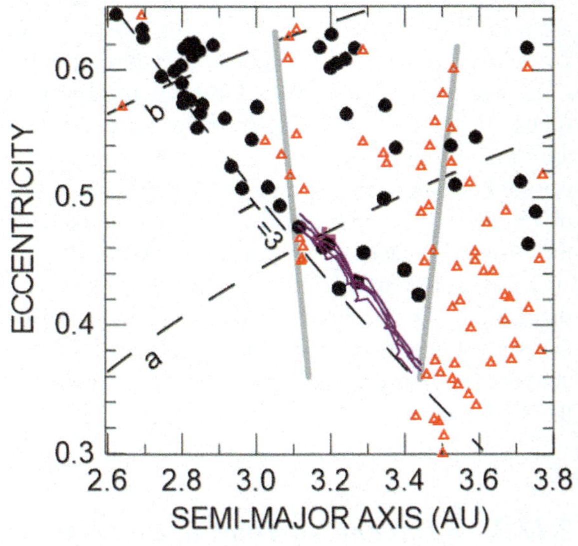

Fig. 2.33 Evolution of the asteroid (5201) Ferraz-Mello in the next 500 years due to the libration (thin purple line; the cross indicates the current position of the asteroid). Asteroids in cometary orbits (balls), and comets of the Jupiter family (red triangles), are also shown. The thick gray lines show the edges of the 2:1 resonance. The dashed lines are the same as in Fig. 2.32 (Reproduced with permission from Carvano et al. (2008) ©ESO)

2.18 Hilda Group

The Hilda group occurs in the 3:2 resonance with Jupiter, where the asteroids have a period of 7.9 years, that is, two thirds of the period of Jupiter. This group includes more than 4000 known asteroids, of which the first to be discovered was (153) Hilda (in 1875, by Johann Palisa) (Fig. 2.34). Unlike the "fragments" present in the Alinda and Hecuba gaps, we find, in this group, some rather large asteroids. Some of them, such as (153) Hilda, (334) Chicago, and (190) Ismene, have diameters of just under 200 km, and another 20 have diameters of 50 km or more. In fact, only the diameters of the larger asteroids are accessible for direct observation. The others are inferred by knowing the intensity of light received from these objects, at various wavelengths. The best determinations come from radiometric observations made in the infrared (mostly by IRAS and MSX satellites (Ryan and Woodward 2010)), where the contribution of the asteroid's thermal emission is large. In the absence of these, in the case of small asteroids, we use the measurement of sunlight reflected by the asteroid and assume that 3.5% of the light they receive from the Sun is reflected (a pessimistic value for most asteroids; the results in these cases are larger than the real object).

The existence of large asteroids in the 3:2 resonance has long been an enigma since important gaps appear in the distribution of asteroids in the locations corresponding to other Jupiter resonances, but not here. These asteroids are not affected in the same way as the others by the phenomena occurring inside resonances. Figure 2.7 shows that, when they align with Jupiter, these asteroids are always near the perihelion. They move close to the stable periodic orbit of the simplified model in which Jupiter is in a circular orbit. We can study them by improving the model, with the motion of the asteroid still perturbed only by Jupiter but with this planet

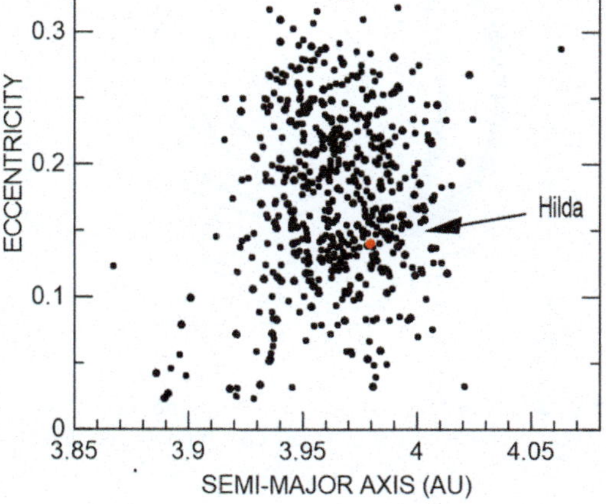

Fig. 2.34 Distribution of semi-major axis and eccentricity of the asteroids in the vicinity of the 3:2 resonance with an absolute magnitude up to 14. The locus of the asteroid (153) Hilda, the paradigm of the group, is indicated

moving in an elliptical orbit. As in the previously studied resonances, the three angular coordinates describing this three-degree-of-freedom motion are chosen in such a way that the symmetries of the equations of motion are preserved. They are:

$$\psi = 2\lambda - 3\lambda',$$
$$\theta = \psi + \varpi, \qquad (2.19)$$
$$Q = \lambda - \lambda'.$$

(see Sect. 2.5). The angle Q varies rapidly, and to reduce the system to only 2 degrees of freedom, it must be eliminated using an averaging or digital filtering method. The angles ψ and θ vary very slowly. If the asteroid is "inside" the resonance, these angles only oscillate with oscillation periods of about 260 years (Ferraz-Mello 1988). The difference between ψ and θ is the longitude of the perihelion, which varies much more slowly. As in the case of the 2:1 resonance (Hecuba gap), from filtered results of numerical integrations of the equations over 1 million years, we can obtain Surfaces of Section. It is worth comparing them with the Surfaces of Section of the 2:1 resonance obtained with the same model (Fig. 2.22).

The first point evident in the figures is the chaoticity in the region where the angle $(\varpi - \varpi_{\mathrm{Jup}})$ oscillates around 180° at high eccentricities (which here are not as high as in the 2:1 resonance). The whole external zone is chaotic. The crescent-shaped curves are blurred, showing that the motions starting on them do not remain over a well-defined tube of orbits for the whole time of integration. In Fig. 2.35a, we show two solutions in the high-eccentricity regime (in crescent shapes). They

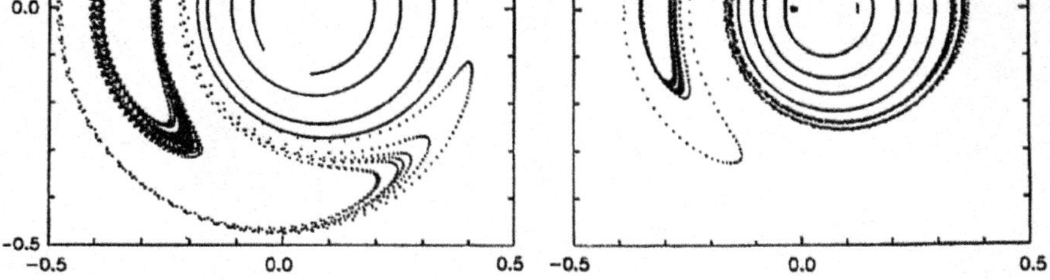

Fig. 2.35 Surfaces of Section of the motion of an asteroid in the 3:2 resonance in two energy levels of the averaged 2D restricted model. The rectangular coordinates are $e.\cos(\varpi - \varpi_{\mathrm{Jup}})$ and $e.\sin(\varpi - \varpi_{\mathrm{Jup}})$ (Reprinted from Ferraz-Mello (1994a) with permission of IAU)

are separated only because the integrations were made over one million years; if we had made them over several million years, the two would surely merge into one. Any asteroid placed in this region will evolve within it and may reach parts farthest from the center of the oscillation, increasing its amplitude; to take the corresponding asteroid out of resonance, we need not wait for the eccentricity to increase greatly and allow the asteroid to approach the inner planets. In the outer curves, the asteroid can come close enough to Jupiter to change its energy and leave the resonance. For that, it is enough that in its oscillation around the periodic orbit corresponding to exact commensurability (see Fig. 2.6), the asteroid accumulates an advance or delay of the order of 60°. In this case, the protection mechanism ceases to work efficiently, the asteroid receives a large impulse from Jupiter and moves away from this region of the asteroid belt. Similar behavior is seen in Fig. 2.35b. In the two Surfaces of Section shown, there are some points already well away, which correspond to the escape of the asteroid from the resonance region. In both figures, we see that the regular inner regions always end before $e = 0.4$ and, in fact, among the observed Hildas, there is none with an eccentricity greater than 0.35.

The LCE of these solutions were calculated in simulations of the evolution of the orbits for 10 million years. The corresponding curves $\lambda(t)$ in the 2D restricted model are shown in Fig. 2.26 and confirm the strong chaoticity in the orbits of higher eccentricity. In the inner kernel, which corresponds to the position of the Hilda asteroids observed in this resonance, the curves $\lambda(t)$ show no inflection, indicating, for this time scale, Lyapunov times greater than 10^6 yr.

Figure 2.36 shows that when we include Saturn and a small inclination (5°) in the orbit of the asteroid, the Lyapunov exponents corresponding to the outermost solutions indicate Lyapunov times on the order of $10^{3.5}$–$10^{4.5}$ years. In contrast, the limits of the λ-functions corresponding to the solutions lying in the innermost part of the section (low eccentricities) lie in the range 10^{-7}–$10^{-5.5}$ yr^{-1} and correspond to Lyapunov times larger than 300,000 years and up to 10 million years.

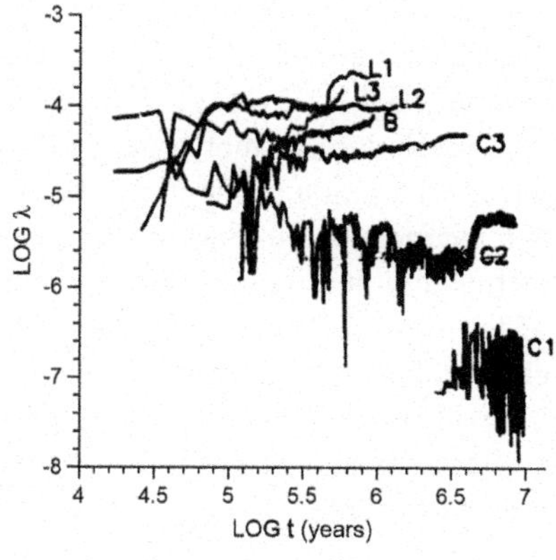

Fig. 2.36 Log-log diagram of the functions $\lambda(t)$ corresponding to several solutions for the motion of an asteroid in the 3:2 resonance, in the 3D model including Jupiter and Saturn and corresponding to the Surface of Section shown in the box included in Fig. 2.26

2.19 Gaps vs. Groups

In this chapter, we discuss how chaos has sculpted the distribution of the asteroids in the main belt. The existence of the Kirkwood gaps has challenged astronomers for one century. As the gaps occur at positions corresponding to orbits with periods commensurable with Jupiter's period, the association of the two facts was immediate. Many books on mechanics indicate that the gaps are due to resonances. But this explanation ignores the fact that there is a group of asteroids located in the 3:2 resonance and two other asteroids located exactly in the 4:3 resonance. This means that the mere occurrence of a resonance is not a sufficient reason to cause a gap. It was only in 1982 that the use of better computers and more elaborate calculation techniques allowed Jack Wisdom to discover the chaotic behavior of the orbits within the 3:1 resonance. But even there, it was clear that the resonance was responsible for the chaos, but chaos alone was not able to create the Alinda gap.

What creates the gap are the asteroids' approaches to the inner planets, which are able to change their orbital energy significantly (see Sect. 2.3) and move the asteroid away from the resonance. If the inner planets did not exist, chaos alone would not be capable of such a change in the innermost asteroids' orbits.

After Wisdom's discovery, it was immediately attempted to extend the same explanation to other gaps in the asteroid distribution. This required another decade. For instance, it was not immediately possible to extend the method used in the 3:1 resonance to the 2:1 resonance, and when it was finally possible to do so, the results showed that in the 2:1 resonance, the chaos is confined to low eccentricities. Intermittencies capable of leading to collisions with other planets do not exist, at least to the extent that the adopted model considers only the perturbations of the asteroidal motion due to Jupiter in a fixed orbit (see Fig. 2.22).

In parallel, the study of the 3:2 resonance, whose dynamics is identical to that of the resonance 2:1, showed similar results. Moreover, a comparison of the Surfaces of Section shows that, in this simpler model, the 3:2 resonance is much less regular (as might be expected, since it is closer to Jupiter and the perturbing effects of the planet are greater). This contrasts with the fact that the 2:1 resonance is nearly empty, and the 3:2 resonance hosts a great number of large asteroids.

The adoption of more complete models is necessary. The inclusion of Saturn and the consideration of the non-coplanarity of the orbits are the next immediate steps. But the inclusion of Saturn and the non-coplanarity increase the number of degrees of freedom and require different methods. The Lyapunov characteristic exponents of the solutions to this problem showed that the 2:1 and 3:2 resonances are both chaotic, but with different degrees of chaoticity. In the 2:1 resonance, the Lyapunov times are always less than 10^5 years, while the 3:2 resonance shows Lyapunov times, at the smallest eccentricities, that can exceed 10^7 years. If we interpret this result with the formula of Lecar et al., we have that orbital transitions are expected in motions within the 2:1 resonance at times shorter than the age of the Solar System. But in the inner parts of the 3:2 resonance, the same transitions are expected at much longer times, up to 10^{10} or 10^{11} years. In this way, the groups led by Sylvio Ferraz-

Mello, at USP, and by Fred Franklin and Myron Lecar, at Harvard, explained why we have a large number of asteroids in the 3:2 resonance, the Hilda group (Ferraz-Mello 1994a; Franklin et al. 1993). By the opposite reason, only small asteroids, likely resulting from collisions close to the 2:1 resonance, exist in the Hecuba gap.

The only difference found between the results for the 2:1 and 3:2 resonance cases is that in the 3:2 resonance, the Lyapunov times of the orbits in the innermost zone of the resonance are of the order of 10^6–10^7 years while, in the 2:1 resonance, they are of the order of 10^4–10^5 years. Hence there was enough time for the escape of asteroids that could originally be in the 2:1 resonance, but not for the same to happen in the 3:2 resonance.

The contrast between the 2:1 and 3:2 asteroidal resonances also appears when they are analyzed following other approaches, as the Frequency Map Analysis introduced in Chap. 3 (see Sect. 3.8). Figure 3.12 shows the frequency maps of these resonances. The difference between the two maps is eloquent. While the map of the 3:2 resonance shows in its central part a large domain where the frequency variation is small (various shades of blue), the corresponding domain in the map of the 2/1 resonance appears fractured in many pieces embedded in a sea of faster frequency variation (yellow). Random walk models say that in the yellow area, the frequency variation may reach 100% at times smaller than the age of the Solar System (Ferraz-Mello et al. 1998b; Nesvorný and Ferraz-Mello 1997). If the blue areas are small, an asteroid placed there will not need much time to escape to the less stable yellow regions. Anyway, the statistical nature of the diffusive processes does not allow us to say that every asteroid in that situation will escape, but the probability of doing so is large. In the same way, an asteroid in the large blue region of the 3/2 resonance may escape, and many may have escaped since the formation of the Solar System, but at a slower pace.

References

Alvarez, W.: T. rex and the Crater of Doom. Princeton University Press, Princeton (2008)

Beaugé, C., Cincotta, P.M.: Shannon entropy applied to the planar restricted three-body problem. Celest. Mech. Dyn. Astron. **131**, 52 (2019)

Benettin, G., Galgani, L., Strelcyn, J.M.: Kolmogorov entropy and numerical experiments. Phys. Rev. A **14**(6), 2338 (1976)

Benettin, G., Galgani, L., Giorgilli, A., Strelcyn, J.M.: Lyapunov characteristic exponents for smooth dynamical systems and for Hamiltonian systems; a method for computing all of them. Part 1: Theory. Meccanica **15**, 9–20 (1980)

Berliner Astronomisches Jahrbuch (Königliches Astronomisches Rechen-Institut zu Berlin, 1919)

Berliner Astronomisches Jahrbuch (Königliches Astronomisches Rechen-Institut zu Berlin, 1932 and 1937).

Born, M.: Vorlesungen uber Atommechanik. Julius Springer, Berlin (1925)

Brouwer, D., Clemence, M.: Methods of Celestial Mechanics. Academic Press, New York (1961)

Brož, M., Vokrouhlický, D., Roig, F., Nesvorný, D., et al.: Yarkovsky origin of the unstable asteroids in the 2/1 mean motion resonance with Jupiter. Mon. Not. R. Astron. Soc. **359**, 1437–1455 (2005)

Carruba, V.: On the age of the Beagle secondary asteroid family. Planet. Space Sci. **166**, 90–100 (2019)

Carruba, V., Nesvorný, D., Aljbaae, S., Domingos, R.C., Huaman, M.: On the oldest asteroid families in the main belt. Mon. Not. R. Astron. Soc. **458**, 3731–3738 (2016)

Carusi, A., Kresák, L., Perozzi, E., Valsecchi, G.B.: Long-Term Evolution of Short-Period Comets. Adam Hilger Ltd., Bristol (1985)

Carvano, J.M., Ferraz-Mello, S., Lazzaro, D.: Physical and dynamical characterization of (2008) Ferraz-Mello, a possible extinct Jupiter family comet. Astron. Astrophys. **489**, 811–817 (2008)

Cincotta, P.M., Giordano, C.M., Silva, R.A., Beaugé, C.: The Shannon entropy: an efficient indicator of dynamical stability. Physica D: Nonlinear Phenom. **417**, 132816 (2021)

Dvorak, R., Lhotka, C.: Celestial Dynamics: Chaoticity and Dynamics of Celestial Systems. Wiley-VCH, Weinheim (2013)

Farinella, P., Froeschlé, C., Froeschlé, Ch., Gonczi, R., et al.: Asteroids falling into the Sun. Nature **371**, 314–317 (1994)

Ferraz-Mello, S.: The high-eccentricity libration of the Hildas. Astron. J. **96**, 400–408 (1988)

Ferraz-Mello, S.: Catalogue of the orbits of resonant asteroids as determined from observations from their discoveries to 1987 (1989). www.astro.iag.usp.br/~dinamica/asteroids_orbits_catalogue/astorb.html

Ferraz-Mello, S.: Kirkwood gaps and resonant groups. Sympos. Int. Astron. Union **160**, 175–188 (1994a)

Ferraz-Mello, S.: Dynamics of the asteroidal 2/1 resonance. Astron. J. **108**, 2330–2337 (1994b)

Ferraz-Mello, S.: The convergence domain of the Laplacian expansion of the disturbing function. Celest. Mech. Dyn. Astron. **58**, 37–52 (1994c)

Ferraz-Mello, S.: Slow and fast diffusion in asteroid-belt resonances: a review. Celest. Mech. Dyn. Astron. **73**, 25–37 (1999)

Ferraz-Mello, S.: Canonical Perturbations Theories. Degenerate Systems and Resonance. Springer, New York (2007)

Ferraz-Mello, S., Klafke, J.C.: A model for the study of very-high-eccentricity asteroidal motion: the 3: 1 resonance. In: Roy, A.E. (ed.) Predictability, Stability and Chaos in N-body Dynamical Systems, pp. 177–184. Plenum, New York (1991)

Ferraz-Mello, S., Tsuchida, M., Klafke, J.C.: On symmetrical planetary corotations. Celest. Mech. Dyn. Astron. **55**, 25–45 (1993)

Ferraz-Mello, S., Klafke, J.C., Michtchenko, T.A., Nesvorný, D.: Chaotic transitions in resonant asteroidal dynamics. Celest. Mech. Dyn. Astron. **64**, 93–105 (1996)

Ferraz-Mello, S., Michtchenko, T.A., Roig, F.: The determinant role of Jupiter's great inequality in the depletion of the Hecuba gap. Astron. J. **116**, 1491–1500 (1998a)

Ferraz-Mello, S., Nesvorný, D., Michtchenko, T.A.: Chaos, diffusion, escape and permanence of resonant asteroids in gaps and groups. In: Lazzaro, D., et al. (eds.) Solar System Formation and Evolution, pp. 65–82. Astron. Soc. Pacific, San Francisco (1998b)

Fouchard, M., Lega, E., Froeschlé, C., Froeschlé, C.: On the relationship between Fast Lyapunov Indicator and periodic orbits for continuous flows. Celest. Mech. Dyn. Astron. **83**, 205–222 (2002)

Franklin, F.A., Lecar, M., Murison, M.: Chaotic orbits and long term stability - an example from asteroids of the Hilda group. Astron. J. **105**, 2336–2343 (1993)

Froeschlé, C.: The Lyapunov characteristic exponents. Applications to celestial mechanics. Celest. Mech. **34**, 95–115 (1984)

Froeschlé, C., Lega, E., Gonczi, P.: Fast Lyapunov Indicators. Application to asteroidal motion. Celest. Mech. Dyn. Astron. **67**, 41–62 (1997)

Froeschlé, C., Guzzo, M., Lega, E.: Graphical evolution of the Arnold web: from order to chaos. Science **289**, 2108–2110 (2000)

Gallardo, T., Ferraz-Mello, S.: Dynamics in the exterior 2: 3 resonance with Neptune. Planet. Space Sci. **46**, 945–965 (1998)

Gladman, B.J., Migliorini, F., Morbidelli, A., Zappalà, V., et al.: Dynamical lifetimes of objects injected into asteroid belt resonances. Science **277**, 197–201 (1997)

Guzzo, M., Lega, E.: Theory and applications of fast Lyapunov indicators to model problems of celestial mechanics. Celest. Mech. Dyn. Astron. **135**, 37 (2023)

Holman, M.J., Wisdom, J.: Dynamical stability in the outer solar system and the delivery of short period comets. Astron. J. **105**, 1987–1999 (1993)

Lecar, M., Franklin, F., Murison, M.: On predicting long-term orbital instability - a relation between the Lyapunov time and sudden orbital transitions. Astron. J. **104**, 1230–1236 (1992a)

Lecar, M., Franklin, F., Soper, P.: On the original distribution of the asteroids: IV. Numerical experiments in the outer asteroid belt. Icarus **96**, 234–250 (1992b)

Levison, H.F., Duncan, M.J.: The gravitational sculpting of the Kuiper belt. Astrophys. J. **406**, L35–L38 (1993)

Michtchenko, T.A., Ferraz-Mello, S.: Comparative study of the asteroidal motion in the 3: 2 and 2: 1 resonances with Jupiter. I. Planar model. Astron. Astrophys. **303**, 945–963 (1995)

Milani, A., Nobili, A.M.: Asteroid 522 Helga is chaotic and stable. In: Dvorak, R., Henrard, J. (eds.) Qualitative and Quantitative Behaviour of Planetary Systems, pp. 323–324. Springer, Dordrecht (1993)

Milani, A., Carpino, M., Hahn, G., Nobili, A.M.: Dynamics of planet-crossing asteroids: classes of orbital behavior: project SPACEGUARD. Icarus **78**, 212–269 (1989)

Moons, M., Morbidelli, A., Migliorini, F.: Dynamical structure of the 2/1 commensurability with Jupiter and the origin of the resonant asteroids. Icarus **135**, 458–468 (1998)

Morbidelli, A., Moons, M.: Numerical evidence on the chaotic nature of the 3/1 mean motion commensurability. Icarus **115**, 60–65 (1995)

Murison, M.A., Lecar, M., Franklin, F.A.: Chaotic motion in the outer asteroid belt and its relation to the age of the Solar System. Astron. J. **108**, 2323–2329 (1994)

Nesvorný, D., Bottke, W.F.: Detection of the Yarkovsky effect for main-belt asteroids. Icarus **170**, 324–342 (2004)

Nesvorný, D., Ferraz-Mello, S.: On the asteroidal population of the first-order Jovian resonances. Icarus **130**, 247–258 (1997)

Perozzi, E., Ferraz-Mello, S. (eds.): Space Manifold Dynamics. Springer, New York (2010)

Roig, F., Nesvorný, D., Ferraz-Mello, S.: Asteroids in the 2:1 resonance with Jupiter: dynamics and size distribution. Mon. Not. R. Astron. Soc. **335**, 417–431. Erratum: id. **336**, 1391–1392 (2002)

Ryan, E.L., Woodward, C.E.: Rectified asteroid albedos and diameters from IRAS and MSX photometry catalogs. Astron. J. **140**, 4 (2010)

Scholl, H., Froeschlé, C.: Asteroidal motion at the 3/1 commensurability. Astron. Astrophys. **33**, 455–458 (1974)

Shevchenko, I.I.: Dynamical Chaos in Planetary Systems. Springer Nature, Chen (2020)

Shevchenko, I.I., Kouprianov, V.V.: On the chaotic rotation of planetary satellites: the Lyapunov spectra and the maximum Lyapunov exponents. Astron. Astrophys. **394**, 663–674 (2002)

Todorović, N., Wu, D., Rosengren, A.J.: The arches of chaos in the Solar System. Sci. Adv. **6**(48), eabd1313 (2022)

Tsiganis, K., Varvoglis, H., Hadjidemetriou, J.D.: Stable chaos in the 12:7 mean motion resonance and its relation to the stickiness effect. Icarus **146**, 240–252 (2000)

Tsiganis, K., Varvoglis, H., Hadjidemetriou, J.D.: Stable chaos versus Kirkwood gaps in the asteroid belt: a comparative study of mean motion resonances. Icarus **159**, 284–299 (2002)

Tsiganis, K., Varvoglis, H., Dvorak, R.: Chaotic diffusion and effective stability of Jupiter Trojans. Celest. Mech. Dyn. Astron. **92**, 71–87 (2005)

Wisdom, J.: The origin of the Kirkwood gaps - a mapping for asteroidal motion near the 3/1 commensurability. Astron. J. **87**, 577–593 (1982)

Wisdom, J.: Chaotic behavior and the origin of the 3/1 Kirkwood gap. Icarus **56**, 51–74 (1983)

Chapter 3
Planetary Systems. Exoplanets

3.1 Chaos in the Solar System

In the previous chapters, we have seen that chaos occurs in many particular situations in the Solar System and is always associated with resonances. Let us now consider chaos in the motion of the planets. Recall that the Solar System is formed by eight planets, each with three degrees of freedom. This is, therefore, a dynamical system that has a total of 24 degrees of freedom.

Several questions can be asked: (1) Is it integrable? (2) Are there KAM tori? (3) How chaotic is it and where? The answer to the first question is immediate: This system is not integrable. (There is no real integrable system with so many degrees of freedom; integrable systems only exist in ultra-simplified models of real phenomena.) The second question is somewhat ambiguous. In general, it is implied that KAM tori are invariant tori remaining in the phase space of a perturbed system, e.g., of a perturbed Keplerian motion (see Chap. 1, Sect. 1.20). It is also common to call KAM tori any invariant tori remaining in the phase space of a nonintegrable system. KAM tori may exist in a system with only the Sun, Jupiter, and an asteroid (Celletti and Chierchia 2007).[1] But, when the system is upgraded and the asteroid is replaced by another planet, the existence of invariant tori is guaranteed only if the masses of the planets do not exceed one thousandth of the mass of Jupiter. But even if this limit is satisfied, what is called perpetual stability does not occur. The existence of these tori is limited to the age of the Solar System (Giorgilli et al. 2009). The existence of invariant tori in the vicinity of planetary motions with the actual

[1] The initial calculations, made more than 50 years ago by Michel Hénon (1966), indicated severe limits for the existence of these tori: the mass of the planet should be smaller than $10^{-48}\mathcal{M}_\odot$ ($\mathcal{M}_\odot$ is the mass of the Sun). At the end of the twentieth century, the Italian mathematicians Alessandra Celletti and Luigi Chierchia, using a new demonstration of Kolmogorov's theorem, succeeded in improving this limit first to $10^{-6}\mathcal{M}_\odot$ and, ten years later, to $10^{-3}\mathcal{M}_\odot$ that is, the mass of Jupiter (Celletti and Chierchia 2007).

S. Ferraz-Mello, *Chaotic Dynamics in Planetary Systems*,
Astronomy and Planetary Sciences, https://doi.org/10.1007/978-3-031-45816-3_3

masses of the planets of the Solar System is far from being proved and may never be proved.

So, after two negative answers, we come to the third question, which is to characterize the chaoticity of the Solar System as it is. Since tidal forces are acting in the Solar System, we cannot rule out the hypothesis that, in the 4.5 billion years of age of the Solar System, the system has evolved to a region of greater stability, close to periodic solutions around which regular motions are expected. Recall the case of Enceladus and Dione (see Sect. 1.18), Saturn's satellites in 2:1 resonance, which are so close to the resonance center that the only way to observe any chaoticity is by drastically changing the parameters of the real problem.

The most immediate way to quantify the chaos in the Solar System is by calculating its Lyapunov exponents (see Sect. 2.10). The first planetary motion for which the Lyapunov characteristic exponents were calculated was that of the dwarf planet Pluto, whose Lyapunov time is 20 million years ($10^{7.3}$ years) (Sussman and Wisdom 1988). At the time that work was done, Pluto was still listed as a planet. But it is, in fact, a very small object—its mass is only 0.2% of the Earth's mass. Its attraction over the other bodies of the Solar System is minuscule, and in most studied problems, its influence can be disregarded. Pluto was classified as a planet when it was discovered by Clyde Tombaugh, in 1930, because of its relatively high brightness. Its surface is icy, with reflectivity (albedo) around 90%, and this led to the belief that it was a much larger object than it really is. Until 1979, its mass was estimated to be about 10% of the Earth's mass, i.e., twice the mass of Mercury, but the discovery of a satellite of Pluto (Charon), and the study of its motion around Pluto, allowed its mass to be precisely determined: 6 times less than the Moon's mass! It is, therefore, an object that seems to be a huge asteroid or, because of its icy and highly reflective surface, a large inactive comet—wandering in the far reaches of the Solar System. Its status as a planet was not justified. In 2006, the International Astronomical Union decided to introduce a new category in the classification of Solar System bodies, the dwarf planets, and included Pluto in this category.

The study of Pluto presents two points that should be emphasized: The first is, as we have already seen, its motion in resonance with Neptune (see Sect. 2.2.4). The period of Neptune's motion around the Sun is 165 years, and that of Pluto is 248 years, i.e., 3/2 of Neptune's period. The second is the time scale of this resonance, which is 30–40 times larger than that of asteroidal resonances because of the long orbital period of the bodies involved. Then, if we study this motion with the same procedures used to study the resonant asteroids, we can go 30–40 times further with the same calculation time. The first long-term integration that could be made of Pluto reached 850 million years, i.e., 1/6 of the age of the Solar System. For the asteroids, with the same computational effort, we would have a time of 20–30 million years; instead of 1/6 of the Solar System's age, we would have half a hundredth of the Solar System's age.

The greatest difficulty in studying the problem of chaos in the complete Solar System comes from the fact that the evolution times are very long. Jerry Sussman and Jack Wisdom initially used two parallel computers built specifically to address this problem with an architecture designed to be efficient in solving the system of equations governing the problem of N planets (Sussman and Wisdom 1992). Considering the 8 planets, they showed that the outer Solar System (Jupiter, Saturn, Uranus, and Neptune) is only weakly chaotic, with Lyapunov times greater than 5 million years ($10^{6.7}$) and the inner Solar System (Mercury, Venus, the Earth, and Mars) is strongly chaotic with Lyapunov times on the order of 10^4–10^5 years.

At Paris' Bureau des Longitudes[2] an institution created 200 years ago with the mission of solving the problem of determining the longitudes, Jacques Laskar approached the problem analytically. The Bureau des Longitudes has a long tradition in the analytical study of planetary motions, and before Laskar's research, a great deal of work was done by Pierre Bretagnon, Jean-Louis Simon, and several collaborators, in which all the algebraic part—algebra of Fourier series with polynomial coefficients—was performed computationally, using algorithms that they built themselves (softwares for algebraic calculation like Mathematica and Maple appeared much later). These theories are of excellent quality and allow us to predict the positions of Mercury, Venus, the Earth (and Moon), Mars, Jupiter, Saturn, Uranus, and Neptune over the next centuries, with great accuracy. The accuracy of the present heliocentric positions of Mercury, Venus, and the Earth is around 0.1 milliarcseconds, and that of the other planets is a few thousandths of one arcsecond (Moisson and Bretagnon 2001). Laskar resumed the analytic theory of motion of these planets and used an averaging method for the elimination of high-frequency (or, as we prefer to call them in Celestial Mechanics, short-period) terms. This dynamical system has initially 24 degrees of freedom. Using an analytical approach and an averaging method, it is possible to reduce it to 16. Instead of working with all the six quantities necessary to fix the position and velocity of each planet, we work only with the four that define the instantaneous orientation and shape of each orbit. Thus we go from a system with some rapidly evolving variables to a new system in which all variables vary very slowly.

Let us see some details of these equations before and after the reduction by the averaging method. To fix the position and velocity of a planet, we need three coordinates: x, y, z, and three components of the velocity v_x, v_y, v_z. Normally, instead of this Cartesian decomposition, we use the three angles that serve to fix the position of the orbit and that of the planet in the orbit—the mean longitude λ, the longitude of the perihelion ϖ and the longitude of the node Ω—and the three momenta associated with these quantities: p_λ, p_ϖ, and p_Ω (which are functions of the semi-major axis a, the eccentricity e, and the inclination i). The quantity λ used to fix the position of the planet in the orbit, varies very quickly. In Mercury, for example, it evolves by 4° per day (the orbital period is 88 days). In the slowest among the 8 considered planets, Neptune, it evolves 2° per year (the orbital period is 165 years). The evolution of the other quantities is much slower. The planes in which the motions take place are almost immobile, and the positions of the planetary

[2] In 1998, the Bureau des Longitudes' Service of Calculations and Celestial Mechanics was transferred to the Paris Observatory, where it became the Institute of Celestial Mechanics and Ephemerides Calculation (IMCCE).

orbits in these planes are almost fixed. The time it takes for the angles ϖ and Ω to complete a revolution is of the order of tens or hundreds of thousands of years. The associated frequencies are at most 0.8° per century. Compared to the mean longitudes λ, there are several orders of magnitude of difference.

Thus, if the integration step for the complete equations must be a few days at most, for the average equations it can be centuries, and the 16 average equations can be integrated in an ordinary computer. This is what Laskar did.

3.2 The Use of the Fourier Transform to Diagnose Chaos

The first chaos diagnosis method we saw in this book was the method of the Surfaces of Section (Sect. 1.12), used for systems with two degrees of freedom, that is, dimension 4 and an integral of energy allowing us to reduce the number of dimensions to 3. When the system has 3 or more degrees of freedom, that method loses its practicality. The result is a figure of 4 (or more) dimensions, and its representation on the paper plane is no longer a Poincaré map in which each point unambiguously determines a solution, but only projections of very difficult interpretation, because, besides the two dimensions represented in the plane of the figure, each point has two, or more, unrepresented dimensions. In some cases, it is possible to draw some conclusions from the roughness of the figure obtained, but, in general, it is impossible to use it for a diagnosis.

For more than two degrees of freedom, other diagnosis methods are necessary, and the most commonly used involves the calculation of the Lyapunov characteristic exponents. Practice shows that in the dynamics of the Solar System, the results obtained present much uncertainty. The discussion concerning the chaotic motion of Helga and the formula of Lecar and co-workers (Sect. 2.15) has made it clear that it is always very difficult to base conclusions on Lyapunov exponents calculated at isolated points in the phase space. For a reliable interpretation, it is necessary to densify the calculation, and this is difficult when it comes to small Lyapunov exponents because of the large computation time required to obtain them.

The next advance in diagnostic techniques was the use of approaches based on Fourier transforms, introduced in studies of dynamics by Donald W. Noid in his doctoral thesis (Noid et al. 1977). As we have already discussed several times, given an integrable system, there are enough conservation laws to define the geometric environment in which the motion takes place. If the system has N degrees of freedom, there is always a transformation of variables that leads to N momenta and N angles, such that the momenta are constant along each solution and the angles are linear functions of time (see Ferraz-Mello 2007, chap. 2). Although this is an often algebraically unfeasible task, in theory it is always possible to transform the variables of an integrable system into momenta and angles with the mentioned properties. Any parameter associated with this system can be written as a function of these variables, and the result is a function periodic in each of the angles. They are quasi-periodic functions of time.

There are several definitions of quasi-periodic functions that we may recall: A quasi-periodic function is a function of the time such that, given an arbitrary ϵ, there exists T such that $|x(t+T) - x(t)| < \epsilon$ for all t (Bohr 1951). The function almost repeats itself identically at every quasi-period T. There is a certain "error" which is given by ϵ. Every such function can be developed into a Fourier series of a certain number of angles given by functions linear in time. If we can represent the function using a finite number of angles, the quasi-periodic function is called conditionally periodic. The adverb is justified by the fact that it will be periodic if one condition is satisfied: the frequencies are two by two commensurable. In this case, the function is periodic and its frequency is the greatest common divisor of the individual frequencies. Example of a quasi-periodic function: any function of type $A\cos\omega_1 t + B\cos\omega_2 t$, with ω_1 and ω_2 incommensurable (if ω_1 and ω_2 are commensurable the function is periodic and the period is given by the least common multiple of the periods $2\pi/\omega_1$ and $2\pi/\omega_2$; if ω_1 and ω_2 are commensurable, there are two integers p and q such that $\omega_1/\omega_2 = p/q$, that is, $p/\omega_1 = q/\omega_2$ which means that after p periods of the wave of frequency ω_1 or q periods of the wave of frequency ω_2, the two waves return to the same phase and the function repeats identically from that time on).

In the case of a system whose solutions are written as quasi-periodic functions of two frequency arguments ω_1 and ω_2 (incommensurable), the Fourier transform of the solutions must appear as a line spectrum. In the example shown in Fig. 3.1a, there are two frequencies and several combinations of them: a main frequency (f), its second and third harmonics ($2f$ and $3f$), a lower frequency (g), its harmonic ($2g$), and several linear combinations of these frequencies ($f-g, f+g, f+2g, 2f+g, 3f-g, \cdots$). In general, if the function that we are considering is a quasi-periodic function, that is, if the system is integrable and has a certain number N of proper frequencies, if we make a transformation from the time domain to the frequency domain by means of a Fourier transform of this function, the result should be a line

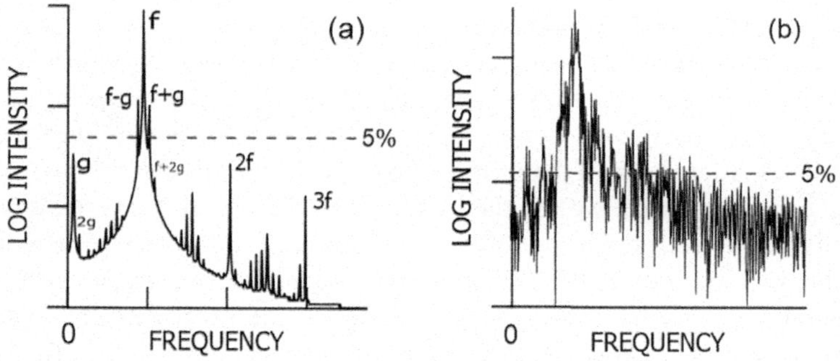

Fig. 3.1 (**a**) Spectrum of a regular solution with two fundamental frequencies f and g. (**b**) Spectrum of a chaotic solution. The scales of intensities are logarithmic to allow details at low intensities to be seen. The dashed lines indicate the height which corresponds to 5% of the highest peak. Frequencies are given in arbitrary units. N.B. The spectral intensity is given by the Fourier transform modulus. (Reproduced from Ferraz-Mello et al. (2005). Copyright 2005 Springer)

spectrum because the only frequencies that may appear are the N proper frequencies and their linear combinations.

The Fourier transform serves, therefore, to support a chaoticity diagnostic method. A spectrum of lines (identifiable with the system's proper frequencies) indicates that a solution is (or seems to be) quasi-periodic, i.e., regular, while a spectrum of bands as shown in Fig. 3.1b indicates that the solution analyzed is not quasi-periodic, i.e., it is chaotic. The precaution shown within parentheses in the previous sentence means that we have here the same practical limitation as in the previous methods (which is a difficulty inherent to the type of phenomenon being studied): A line spectrum can be either the spectrum of an integrable system, or the spectrum of a nonintegrable system whose solutions behave like those of an integrable system during the time interval used to construct the spectrum.

We also have the opposite possibility of obtaining a spectrum that does not show lines just because the sampling interval of the analyzed function is not long enough. The reason for this last difficulty is that the analyses are made with samples of the functions with a finite number of points. Therefore, the integral appearing in the definition of the Fourier transform (which becomes a sum of $-\infty$ to ∞ in the discrete version) is, in fact, replaced by a finite integral (or sum), covering only the instants of the time interval in which the function was sampled. We thus have a finite interval and a finite number of points. The results are not the same as when the function is known from $-\infty$ to ∞. In the resulting spectra, the lines are replaced by peaks with some opening at the base. It is easy to see what happens in this case sampling points of a simple sinusoid on a finite interval and using the usual formulas to calculate the Fourier transform of the time series so constructed. At first sight, a spectrum may look like a band spectrum, meaning that the orbit being studied is chaotic and does not have clearly defined frequencies while, in fact, the system has just not been sampled long enough to separate well the various lines of the spectrum. These difficulties are solved by sampling for a longer time interval.

In some of the techniques proposed by Jacques Laskar and described in this chapter, determining the frequencies of the main peaks of the spectrum plays an essential role. In these techniques, the first thing we do is to construct, from the outputs of numerical integration, series of values taken in instants separated by a constant time interval chosen in such a way that the series gives a good representation of the studied motion. Next, we compute the Fourier transform of the series so constructed. The Fourier transform is a computationally long operation. To save computing time, we use the Fast Fourier Transform (FFT) algorithm. For this purpose, the series with the results of the integration must have a number of terms that is a power of 2. The FFT uses this particularity to avoid the repetition of calculation segments that remain the same when the number of points in the series is a power of 2. But a Fourier transform of a series with, say, $2^{10} = 1024$ points, is a series with the value of the power spectrum in 1024 points. If we are entering information given by 1024 numbers, as the Fourier transform does not create or destroy information, we can only have 1024 numbers in the output. This precision is, in general, insufficient for the type of study that we want to do. Almost certainly, the frequency which we want to determine is not one of the 1024 frequencies of the output and lies between two of them. But once we have located the interval where the peak with the desired frequency is, we can use a finer spectral analysis and interpolations to find the exact frequency that corresponds to the maximum value of the Fourier transform intensity. The ideas to achieve this are simple. We know that in the spectrum of a finite number of points, a spectral line has the form of a *sinc* function. In its central part, this function can be approximated by a parabola. It is enough, therefore, to construct a tightly closed grid with a usual spectral analysis algorithm for time series, then pass a parabola through the highest points and determine its vertex. With this, it is possible to determine the frequency of the selected peak with the necessary precision (Laskar et al. 1992). This method can be sophisticated and routines created, for example, to determine the frequencies corresponding to the maxima in the critical cases where two frequencies are very close to each other. In this case, two peaks, which in other circumstances would appear as isolated, are composed. The result is a curve much more complicated than the mere sum of the two peaks, since spectral intensities are obtained from the modulus of complex quantities.

3.3 Chaos Around the Giant Planets. Dynamical Maps

The Fourier transforms can be used to construct dynamical maps of regions of the phase space in order to identify local structures responsible for the chaoticity of the solutions. These maps are built on two-dimensional grids of initial conditions, in which two of the parameters are varied at time t_0 while keeping fixed all the others. Then, solutions of the system are built for each of the initial conditions of the grid through numerical integrations. These solutions are then analyzed using a chaoticity indicator that somehow quantifies the behavior of each solution. An important point for the interpretation of the results is the correct identification of the proper frequencies of the system, which can be done by comparing the solutions to previous estimates obtained with an approximate analytical model. Methods based on Fourier transforms make easy the identification of the frequencies associated with chaotic motions.

An alternative indicator of chaoticity is the spectral number introduced by Tatiana Michtchenko (Michtchenko and Ferraz-Mello 2001). It is obtained by counting the number of Fourier Transform peaks above a pre-established noise level. For instance, if we define the threshold level as 5% of the intensity of the highest peak in the spectrum, the spectral number associated with the spectrum shown in Fig. 3.1a is $N = 3$. In contrast, the spectral number associated with the spectrum shown in Fig. 3.1b is a large number difficult to determine from the figure; we would need access to the table used to construct the figure and count there the number of peaks above the noise threshold level. In practice, there is not much point in making the number of peaks more precise in these cases, and the counting is stopped when a value is reached beyond which the accuracy of this number is no longer significant.

In general, we set this maximum value to $N = 80$, or $N = 100$, and associate this value to all spectra with a number of peaks larger than this limit.

Dynamical maps constructed in this way have been used to learn about the surroundings of the giant planets of the Solar System (Jupiter, Saturn, Uranus, and Neptune). They are shown and discussed below. In each of them, the initial conditions of the simulations were obtained by varying the initial semi-major axis and eccentricity of the planet within the limits shown in the figures and using the current values for the other orbital elements of the considered planet and those of the other planets. The simulations take into account all the interactions between the Sun and the four giant planets. The interactions with the inner planets (Mercury, Venus, Earth, and Mars) were not considered because the masses of these planets are much smaller than those of the giant planets.

The first map (Fig. 3.2) shows the surroundings of Jupiter. It shows a wide band of chaotic motion at the position corresponding to the 5:2 resonance with Saturn (marked with (a) on the map). Orbits initiated in this region have large perturbations, mainly in eccentricity. However, at low initial eccentricities, in a large domain (white on the map), the orbits remain relatively stable for up to 800 million years (the length of some of the simulations reported in Michtchenko and Ferraz-Mello (2001)).

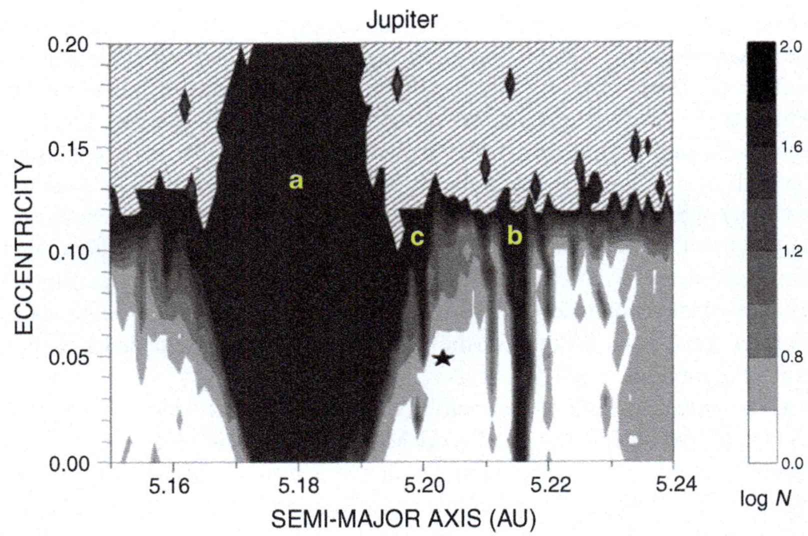

Fig. 3.2 Dynamical map of the Jupiter neighborhood. The light regions are of regular motion and the dark regions are of chaotic motion. The chaoticity measured by the spectral number N is represented according to the scale next to the figure. The hatched areas correspond to initial conditions leading to planetary collisions in less than 1.5 million years. The current position of the planet is indicated by a star. The main domains of initial conditions leading to chaotic motions associated with resonances are indicated by letters and discussed in the text. (Reproduced from Michtchenko and Ferraz-Mello (2001) with permission of the AAS.)

The behavior resembles that seen in the 3:1 asteroidal resonance: the asteroids had large excursions in eccentricity due to the resonance but returned to lower values. The clearing of the 3:1 asteroidal resonance only occurs because, at times of high eccentricity, the asteroid can come very close to Mars or another of the inner planets and then undergo a large variation in its orbital energy and move away from the resonance. In the absence of these collisions, nothing more happens than the existence of occasions when the eccentricity gets high.

The major consequence of an eventual approach of the Jupiter-Saturn pair to the 5:2 resonance is not in the motion of Jupiter or Saturn, but in the motion of the other Solar System bodies. A large eccentricity in Jupiter's orbit induces large eccentricities in the asteroidal orbits, and asteroids in orbits with large eccentricities have encounters with planets capable of expelling them from the domain in which they find themselves. This is what is seen in the simulations of the rapid planetary migrations that took place in the early Solar System when resonances are crossed. For instance, in the model of the formation of the Solar System known as the "Nice model" (Gomes et al. 2005; Tsiganis et al. 2005a), the crossing of the 2:1 resonance of Jupiter with Saturn (in which case the orbital period of Saturn is twice that of Jupiter) causes a disturbance in the disk of planetesimals, greatly increasing their orbital eccentricities and favoring their collisions with planets. As a result of this cataclysm, the disk of debris left over from the formation of the planets disappears. In the Nice model, the crossing of the 2:1 resonance occurs at about 700 million years after the formation of the planets and, thus, coincides with the period known as Late Heavy Bombardment (LHB), a period in which lunar craters were formed by large impacts, as determined from samples brought back from the Moon by the Apollo astronauts. However, later studies indicate that this period of intense bombardment occurred during a much longer time and much earlier, tens of million years after the formation of the planets (Mojzsis et al. 2019). The instability in this case would have been caused by multiple resonance crossings of large planetary cores that formed beyond the orbit of Saturn (de Sousa Ribeiro et al. 2020).

In parallel with the large chaotic band associated with the 5:2 resonance of Jupiter with Saturn, the dynamical map in Fig. 3.2 shows several lines of chaotic motions that can be associated with frequencies not only of Jupiter and Saturn but also of Uranus. The line identified with the letter b, for instance, is a 3-planet resonance in which the frequency of the angle associated with the 5:2 resonance of Jupiter with Saturn is equal to half that of the angle associated with the 3:1 quasi-resonance between Saturn and Uranus. The line identified with the letter c is similar, with the frequency of the angle associated with the 3:1 quasi-resonance between Saturn and Uranus being equal to twice the frequency associated with the 5:2 resonance of Jupiter and Saturn. The two lines next to line b, one on each side, are even more complex, also involving the frequency of the angle associated with the 2:1 quasi-resonance between Uranus and Neptune; these are resonances involving the frequencies of the 4 planets.

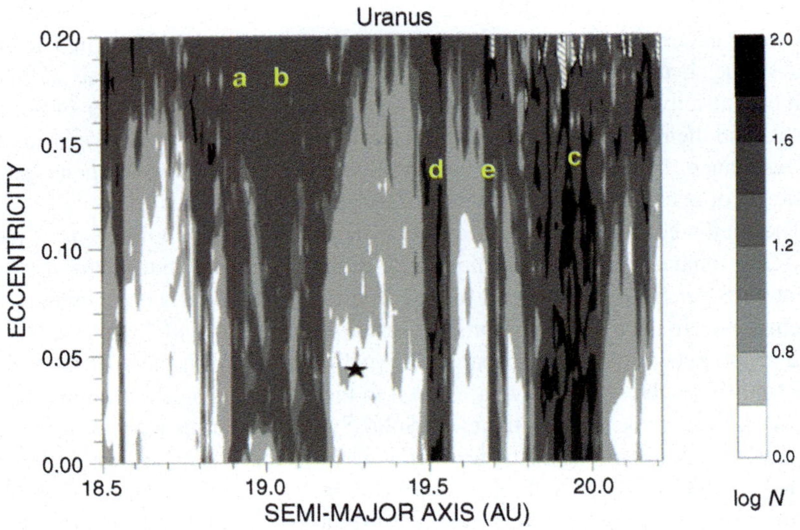

Fig. 3.3 Dynamical map of the surroundings of Uranus using the same convention as in Fig. 3.2. The small hatched areas on top of the figure correspond to initial conditions leading to planetary collisions in less than 3 million years. (Reproduced from Michtchenko and Ferraz-Mello (2001) with permission of the AAS.)

The second map (Fig. 3.3) shows the surroundings of Uranus. Because of the situation of Uranus in the planetary system, this dynamical map shows a large number of lines of chaotic activity parallel to one another. The most important one is that corresponding to the 2:1 resonance with Neptune (marked with a on the map), which blends with the line corresponding to the 7:1 resonance with Jupiter (marked with b on the map). Another simple resonance (that is, involving only two planets) appearing in the map corresponds to the 3:1 resonance with Saturn (marked with c on the map). In addition, we observe many beat lines from these resonances. For instance, the lines marked with d and e are from beatings of the frequency of the angle associated with the 3:1 resonance with Saturn and that of the angle associated with the 5:2 resonance between Jupiter and Saturn.

The last two maps (Fig. 3.4) show the neighborhoods of Saturn and Neptune. The orbit of Saturn is close to that of Jupiter, and for this reason, the map of Saturn's surroundings is dominated by chaos due to the 5:2 resonance of Jupiter and Saturn. The other visible lines are greatly deformed by their proximity to the 5:2 resonance. For instance, the line marked b on the map corresponds to the beat of the frequency associated with the 3:1 resonance with Saturn with that associated with the 5:2 resonance between Jupiter and Saturn. The map of Neptune's surroundings shows only the chaotic domain associated with the 2:1 resonance of Neptune and Uranus.

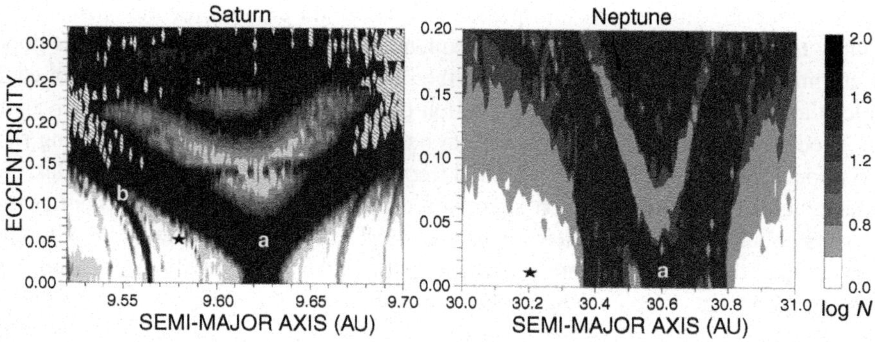

Fig. 3.4 Dynamical maps of the surroundings of Saturn and Neptune using the same conventions as in Fig. 3.2. (Reproduced from Michtchenko and Ferraz-Mello (2001) with permission of the AAS.)

3.4 Frequency Analysis of Weakly Chaotic Systems

If a solution is strongly chaotic, the Fourier transform can indicate this stochasticity without any doubts. But the technical problems described in Sect. 3.2 become crucial in the case of only weakly chaotic solutions. A nonintegrable system that differs little from an integrable system (i.e. being only weakly non-integrable) can, for short times, be confused with an integrable system (the solution bundles of both are very close without being possible to distinguish them numerically). In short times, their solutions evolve in an approximately quasi-periodic way. But only for short times. In this case, the Fourier transform of a solution extending over a short interval result in a spectrum of lines, as in integrable systems.

Based on this fact, Jacques Laskar has elaborated several strategies for the frequency analysis of systems with many degrees of freedom and weak chaoticity, which go beyond the mere diagnostic of chaoticity and allow one to obtain global characteristics of the chaotic behavior of the solutions of a system.

Let us start with the first of these techniques and its application to the Solar System. Initially, we construct the solution over a sufficiently long time interval using numerical integration. At the end of the integration, we select one or more parameters and sample them by taking their value at each time interval Δt. The next operation is to obtain the Fourier transform of the time series built in this way. This Fourier transform allow us to verify the higher or lower chaoticity of the solution by the higher or lower complexity of the spectrum. To improve the diagnosis, instead of building the Fourier transform of the entire time series resulting from the integration, we divide the time series into pieces and perform the Fourier transform and the frequency analysis of each piece separately. This division is critical. The pieces should be short so that the Fourier transform gives a line spectrum. But if the pieces are too small, the lines will broaden and blend together, so that it is impossible to determine the frequencies to which they correspond. Therefore, the first operation is to divide the time series into pieces that are small enough for the spectrum to allow the identification of the main frequencies but large enough to allow a good determination of these frequencies. We then calculate the Fourier transform of each piece and make an accurate determination of the main frequencies. If the division is good, in each of the pieces the system behaves in a very regular way giving rise to a spectrum of lines with well-defined frequencies. If the system were regular, the frequencies determined in each slice would be the same, but as it is chaotic, the results obtained from each slice are different. We can then make graphs showing the variation of the frequency, and from them, to know the characteristics of the motion. The motion takes place in a chaotic layer, and the variation of the frequencies allows us to visualize the width of that chaotic layer. This is a more complete information than the simple chaoticity diagnostic. For this purpose, additional information is needed; we can infer what happens to the momentum corresponding to the changing frequency and try to localize the chaotic layer in terms of the momenta, but this will depend on having a good analytic model of the problem under study. If the frequency limits are ω_a and ω_b, we need the analytical model to identify the values p_a and p_b of the corresponding momenta.

The first application of this technique was made by Laskar in the study of the chaoticity of the planets of the Solar System, a system which, as already discussed, even when eliminating the high-frequency terms, has 16 degrees of freedom. These motions are chaotic, but not enough to prevent the basic frequencies from continuing to exist. They exist but vary throughout the solution.

In the case of the planets, the important frequencies are due to the motions of the nodes and the perihelions (the frequencies of the orbital motions were eliminated in the averaging process). Because of the strong interaction between the planets, it is not possible to separate the various frequencies: they appear with greater or lesser intensity in the spectra of all the planets. The proper frequencies that occur in the motion of one planet also occur in the motion of the other. But it is possible to identify which is the most important in the motion of each planet. Thus, g_1 is associated with the longitude of Mercury's perihelion, g_2 with that of Venus, g_3 with that of the Earth, etc. In the same way, the proper frequencies associated with the longitudes of the ascending nodes of the orbital planes are also determined (Laskar 1990). Figure 3.5 shows the results obtained for some of the proper perihelions in an interval of 200 million years divided into overlapping segments of 20 million years each.

3.5 The Interplanetary Spacings

Frequency analysis confirms the chaoticity of the inner Solar System, already characterized by a Lyapunov time of 5 million years (Laskar 1990). Can we say that this system is stable? To answer this question, Laskar performed several integrations of the average equations spanning 25 billion years: 15 into the future and 10 into the past. This requires explanation: What sense is there in integrating into the past

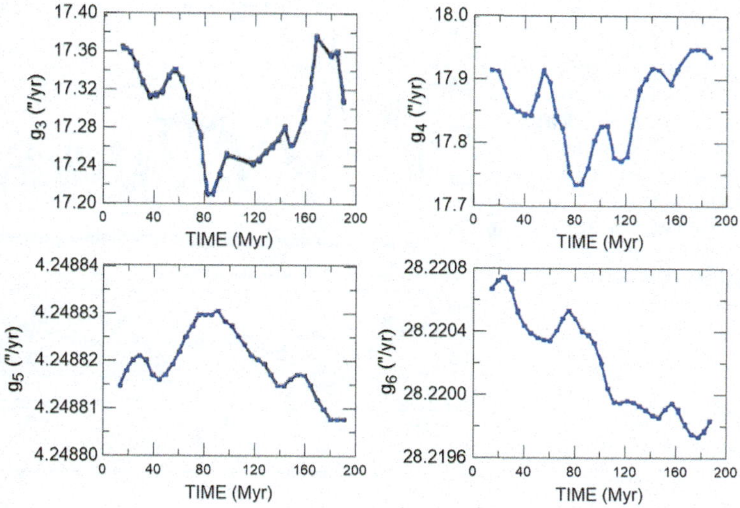

Fig. 3.5 Evolution of the proper frequencies associated with the perihelions of Earth (g_3), Mars (g_4), Jupiter (g_5), and Saturn (g_6) in arcseconds per year. Time in million years. (Adapted from Laskar (1990). Copyright 1990 by Elsevier.)

for a time twice as long as the age of the Solar System? But, even forgetting this fact, what sense does it make to integrate the equations of motion for such a long time? In fact, no integration of this system is valid for a time longer than 50 million years. Simulations over longer times have the sole function of letting a solution wander through phase space, in the hope that, by passing through some of its most recondite domains, it reveals some of its secrets. One can also justify excessively long integrations by a conjecture known as the "shadow principle": because of the chaotic nature of the system, some solution exists in the vicinity of the solution obtained, even if it is not the solution that strictly corresponds to the initial condition set (this conjecture is based on the questions of Symbolic Mechanics presented in Sect. 1.18).

The results are shown in Fig. 3.6. The following points are worth mentioning: (1) the huge jump in Mercury's eccentricity and inclination (reaching $e = 0.52$ and $i = 20°$); (2) the small and identical variations of the orbits of the Earth and Venus; (3) the absence of variations in the orbits of the outer planets: Jupiter, Saturn, Uranus, and Neptune.

A simple calculation shows that maintaining their average distances from the Sun (which could only be significantly changed by a swing-by), the maximum eccentricities reached in these simulations are not enough to allow the planetary orbits to cross one another. But by drawing the zones defined by the maximum and minimum distance of each planet from the Sun, we see that the edges of their motion zones are not very far from touching. These results show that the Solar System can be considered as being "stable" for times much longer than the age of the Solar System. Figure 3.7 shows that the inner Solar System is "full". There is no room

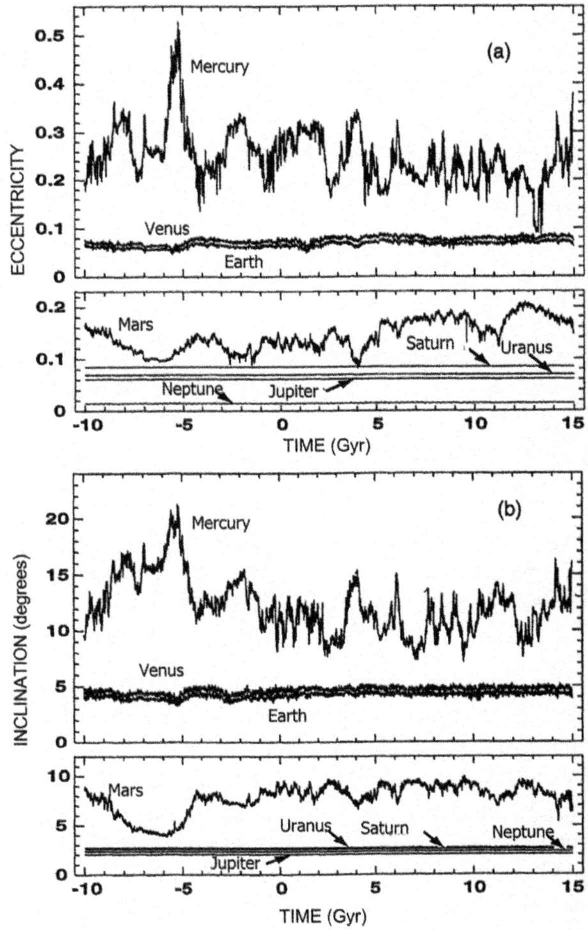

Fig. 3.6 Numerical integration of the averaged equations of motion of the planets of the Solar System for 25 billion years. For each planet, the graphs show the maximum values reached by the eccentricity and inclination at intervals of 10 million years. (Reprinted from Laskar (1996) with permission of IAU.)

for other orbits in the interplanetary spacings of the inner Solar System. Any other object placed in orbit between the inner edge of the asteroid belt and the Sun will inevitably cross the orbits of the present four inner planets, with the occurrence of close encounters, or even collisions, and profound changes in the system. A fifth planet in this region would imply instability of the system and certainty of some collisional event capable of significantly changing the history of the system. It is of interest to remember that the most accepted theories of the origin of the Moon involves the collision of the Earth with a planet smaller than Mars: when approaching the Earth, this planet (which was called Theia) would have collapsed because of the strong tides caused on it by the Earth, its heavy core would have collided with the Earth, and the lighter parts of its mantle would have spread out in the vicinity of the Earth, forming a ring of matter that later merged to form the Moon (this would explain the low density of the Moon, 3.36 g/cm^3, and the fact that the Earth is the planet with the highest density: 5.52 g/cm^3).

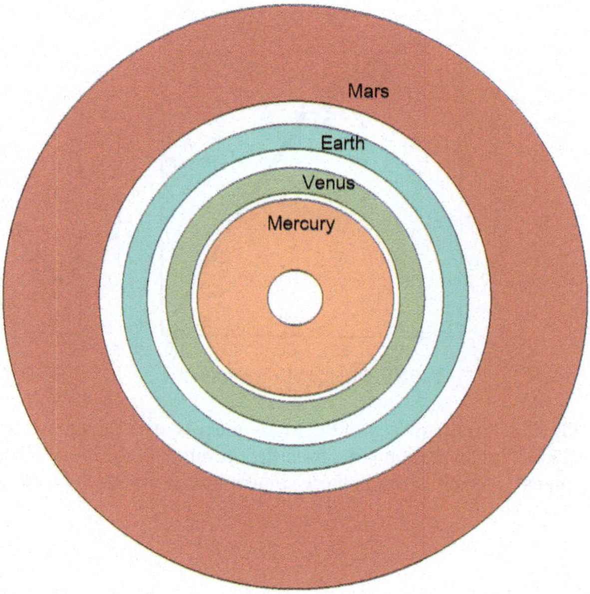

Fig. 3.7 Zones visited by each of the inner planets of the Solar System during 5 billion years

Some bolder extrapolations indicate that the eccentricity of Mercury could grow even larger than the 0.52 of the figures shown and that, in some billion years Venus and Mercury may collide; if confirmed, this result would mean that the stability of the inner Solar System is not perennial, but is limited to the next billions of years.

The outer Solar System is much more stable. But also there, simulations show that any object placed between the orbits of Jupiter and Neptune quickly tends to have a close encounter with one of the four planets. It would be difficult to accommodate a larger number of giant planets in the same space occupied by the four current giant planets. But this does not mean that it is impossible. Simulations made to understand how the Solar System was formed often lead to solutions with a different number of giant planets, in arrangements apparently stable.

3.6 The Rotations of the Earth and Mars

Laskar used frequency analysis in several different ways. Laskar's first technique was the one used to study the chaoticity of the Solar System by determining the time-variation of the proper frequencies (Sect. 3.4). The following ones went on to look not only at the variation of frequencies in consecutive segments of the solution, but did so with many solutions starting at many points in a domain of phase space. One of the occasions where this technique was applied was in the study of the rotations of Mars and the Earth (Laskar and Robutel 1993).

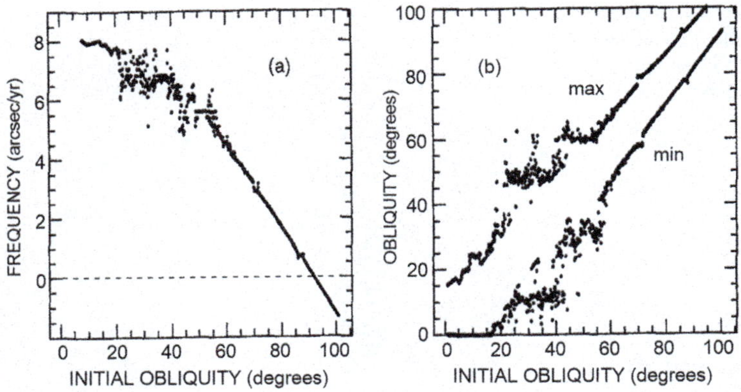

Fig. 3.8 (**a**) Frequency of the Mars' precession obtained by analyzing simulations of the rotation of Mars for 56 million years for different values of the initial obliquity. (**b**) Maximum and minimum values of the obliquity in these simulations. (Reproduced from Laskar (1996). Copyright 1996 by Springer

Several studies of the rotation of Mars have shown that it is chaotic: it is known that the obliquity of Mars varies between 10° and 50° because of planetary perturbations.[3] Let us remember that, when a planet rotates around an axis, this axis does not remain fixed in space but moves continuously drawing a cone—like the axis of a spinning top. This movement is called precession and has a very long period. In the case of Earth, the current period of precession is 25,725 years. Mars' precession is somewhat slower, with a period of about 160,000 years. To understand the chaotic rotational dynamics of Mars, Jacques Laskar and Philippe Robutel used frequency analysis to see how the frequency of the planet's precession varies. To do this, they integrated the equations of Mars' rotational motion and simulated the motion for 56 million years for a large number of initial conditions. They kept the initial values of all variables the same but one: the obliquity of Mars. The solutions were then analyzed and, in each case, the frequency of the precession was determined. For values of the initial obliquity above 60°, the results are arranged on a well-drawn curve, showing that the frequency of precession decreases when the obliquity increases and becomes negative when the obliquity exceeds 90°. But between 0° and 60°, the values obtained for the precession varies from one integration to the next in an irregular way indicating the existence of a strong chaotic zone (Fig. 3.8) A complementary figure shows the corresponding variations of the obliquity of Mars (the minimum and maximum values of the obliquity).

[3] The obliquity of a planet is the angle (dihedral) formed by the plane of the planet's equator and the plane of its motion around the Sun. Depending on whether the planet is rotating in a direct or retrograde direction, i.e., with the same direction or opposite direction with respect to the orbital motion of the planet around the Sun, we take the angle less than or greater than 90°, respectively.

The same study can be done for the other planets. The rotations of Mercury and of Venus are peculiar: Mercury dissipated its rotational energy in the friction of its internal parts constantly deformed by the tides caused by the strong solar attraction and ended up with its rotation captured in a 2/3 commensurability with its orbital period (see Ferraz-Mello 2015). Venus was also affected by solar tides, but to a lesser extent, and had its state of retrograde rotation with a period of 253 days dictated by the dynamics of the interaction of its core and mantle. The rotation of a planet affects its flattening (which results from the balance of gravitational cohesion and centrifugal forces); Venus and Mercury have very slow rotations and small flattenings. Their precessions are close to 0 and are regular.

In the case of the Earth, the problem is complicated by the presence of the Moon. Repeating for this system the same calculations described for Mars, we obtain (Fig. 3.9a–b) fairly regular motions, except for a range of obliquities between the

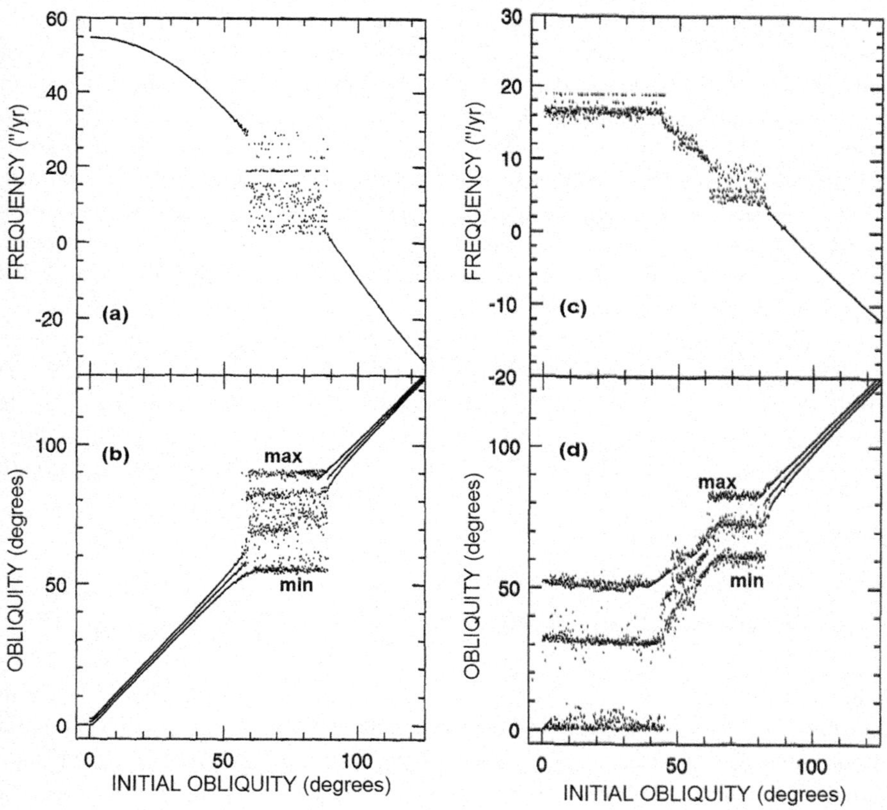

Fig. 3.9 (**a**) Frequency of the Earth's precession obtained by analyzing simulations of the Earth's rotation for 18 million years for different values of the initial obliquity. (**b**) Average, maximum, and minimum values of the obliquity in these simulations. (**c–d**) The same parameters as in (**a–b**) obtained with a simplified model that considers only the torque of the forces due to the Sun and ignores that of the forces due to the Moon. (Reproduced from Laskar (1996). Copyright 1996 by Springer.)

limits 60° and 90°, far away from the present 23°. But it is interesting to note that, excluding the Moon, the result would be very similar to that of Mars with a chaotic zone between 0° and 90° (Fig. 3.9c–d). This allows us to say that the regularity of the climate on Earth exists only because of the Moon. Were it not for the existence of the Moon around the Earth, the Earth would have a chaotic rotational regime in which a few million years would separate "glacial" and "torrid" epochs, and on some occasions (when the obliquity is around 90°), the polar night and day would reach almost the whole Earth, freezing and then roasting its higher latitude regions. Such a thermally convulsed environment might not have allowed the development of more evolved forms of life.

The mechanism by which the Moon prevents a chaotic motion of the Earth's axis is the influence of the Moon on the frequency of precession of the Earth's axis of rotation. This frequency is induced by the torque applied to the Earth, and the contribution of the Moon's attraction is about twice that of the Sun. This fact makes the frequency of the Earth's precession much higher than that of Mars, and is at the origin of such different behaviors. If we remember that the Moon is slowly moving away from the Earth due to the tides caused on Earth by the Moon, we can imagine that with the Moon moving away, the contribution of the lunar attraction loses importance, and the rotation of the Earth becomes chaotic. This will indeed happen! When the Moon moves away from the Earth by a little more than half of its present distance, the frequency of the precession will drop to half its current value, and the motion of the Earth's axis will become chaotic, allowing the Earth's obliquity to occasionally reach values close to 90°. But a lot of time will be needed before that happens. Probably more than a billion years (Correia and Laskar 2010; Neron de Surgy and Laskar 1997).

As for the large planets, their rotations are regular. The reason for this can be seen in Figs. 3.2, 3.3, and 3.4 where is shown that the large planets are in domains of these maps where $N \leq 3$, which means that their Fourier spectra (FFT) include few lines with significant amplitudes. The possibilities of overlapping and beating between the frequencies of precession and the frequencies of the Fourier spectra are small.

3.7 Frequency Analysis on Dense Grids. Arnold Web

The third strategy of using Frequency Analysis, several times explored by Laskar in systems with 3 or more degrees of freedom, was that of the follow-up of several main frequencies (instead of a single frequency) and the use of frequency vs. frequency graphs. Or rather, calling the independent main frequencies f_1, f_2, f_3, the graphs of the ratios $\rho_1 = f_1/f_3, \rho_2 = f_2/f_3$. These graphs reveal the entire resonance web (Arnold web) of a given phenomenon and the places where their influence is the greatest.

Let us take as an example the graphs obtained in the problem of the motion of electrons in the storage ring of an accelerator. The graph in Fig. 3.10a was

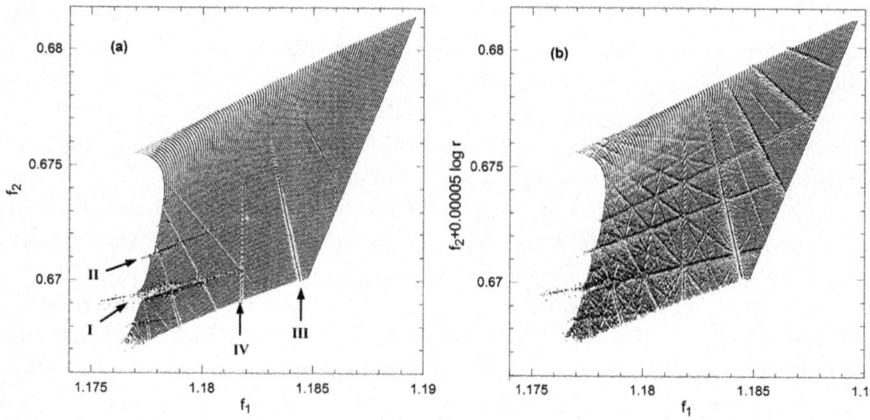

Fig. 3.10 (**a**) Image in the frequency plane obtained using frequency analysis to determine the proper frequencies f_1, f_2 in solutions generated from a grid of initial conditions. The lines indicated with I, II, III, and IV correspond to the resonances identified in the text. (**b**) Transport ratio in the frequency plane. The transport ratio is visualized by adding $0.00005 \log r$ to the frequency f_2 at the points of the figure (**a**) (r is the variation of one of the frequencies. See text). (Reproduced with permission from Dumas and Laskar (1993). Copyright 1993 by the American Physical Society.)

obtained exactly as indicated above. The motion of the electrons is modeled by a Hamiltonian system with two degrees of freedom, not autonomous. The first two frequencies are the proper frequencies of the system, and the third frequency corresponds to the extension of the phase space with the inclusion of a third degree of freedom, in which time is the angular variable. The third frequency is, therefore, equal to 1. The solutions are calculated by taking as initial conditions fixed values for the angles (for instance $\phi_1 = \phi_2 = 0$) and values of the momenta (or actions) I_1, I_2 in a rectangular grid, as dense as possible. If the system is regular, the lines corresponding to a grid line will be continuous with only a slight deformation. This is approximately the case except in the vicinity of some lines where abrupt distortions occur. These distortions correspond to resonances in which the proper frequencies satisfy a commensurability relation of the type $k_1 f_1 + k_2 f_2 + k_3 f_3 = 0$ where the k_i are integers. As an example, we identify some of these lines: Line I in the figure corresponds to the resonance $2f_1 - 8f_2 + 3 = 0$ (recall that $f_3 = 1$). Line II corresponds to $2f_1 - 5f_2 + 1 = 0$. Line III corresponds to $9f_1 + 2f_2 - 12 = 0$ and, finally, vertical line IV corresponds to $11f_1 - 13 = 0$ (the frequency f_1 is a constant commensurable with $f_3 = 1$). The numbers appearing in these equations are not always low and show that even resonances involving not-so-low numbers matter when studying the chaoticity of a dynamical system.[4]

[4] The order of one resonance is the modulus of $\sum k_i$. The four selected lines are of orders 3, 2, 1, and 2, respectively.

Figure 3.10b is slightly different in its design. In it, not only do we compute the main frequencies in a given time interval, but also in a second time interval following the first one, and measure the frequency changes between the first and the second interval. On the vertical axis of the new figure, instead of f_2, we put $f_2 + 0.00005\log(r)$, where r is the variation of the frequency that had the greatest variations between the first and the second interval.

The emphasis that this combination gives to the resonance lines is simply amazing. This set of lines forming a web is called the "Arnold web". The enhanced lines correspond to the commensurabilities $k_1 f_1 + k_2 f_2 + k_3 f_3 = 0$ up to a certain order. In Fig. 3.10b, this web emerges from the computation of the fundamental frequencies and their variations and shows the commensurabilities and their importance. Note that the disturbance visible in the graph, closer to the highest vertex in the figure (vertex that corresponds to the center of the ring), is associated with the resonance $3f_1 + 8f_2 - 9 = 0$. This resonance belongs to a beam with lines coming out of the radiant (13/11,15/22) located at the upper edge of the figure.

The result shown in Fig. 3.10b can be presented in another way. Instead of a coordinate plane f_1, f_2, we take the grid of momenta I_1, I_2. If we plot the values of I_1 and I_2 we have only the grid of initial values without any information on the solution. But if we add an amount proportional to $\log(r)$ on one of the axes, we get the analogous of the last figure shown, but in the space of momenta instead of the space of frequencies. Figure 3.11 is the graph of the points $I_1, I_2 + 0.04\log(r)$; it shows the Arnold web in the momenta instead of the frequencies. The results are equivalent to those obtained in the plane of the frequencies, but the use of moments makes it easier to identify the characteristic parameters of the system under study.

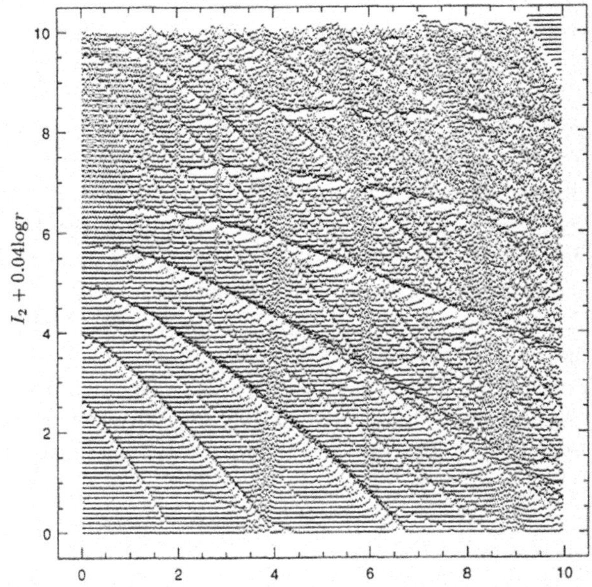

Fig. 3.11 Transport ratio in the plane of moments. The variables of the graph are I_1, I_2. The transport ratio corresponding to the initial conditions (I_1, I_2) is visualized by adding $0.04\log r$ to the momentum I_2. (Reprinted with permission from Dumas and Laskar (1993). Copyright 1993 by the American Physical Society.)

In the new graph, the center of the ring is the point at which the moments I_1, I_2 are zero.

In planetary dynamics problems, the moments are associated with the geometric elements of the orbit, semi-major axis and eccentricity, which are generally used instead of the momenta. Another modification is that the plotted quantity is directly the transport ratio r.

3.8 Other Strategies in Frequency Analysis. Resonant Asteroids

In the chapter on resonant asteroids, the last result we arrived at was the one obtained with Lyapunov times and the Lecar et al.'s rule (Sects. 2.14 and 2.15). According to these results, the calculated Lyapunov times indicate the possibility of eventualities (sudden orbital transitions) between 10^7 and 10^9 years, in the 2:1 resonance, and between 10^{10} and 10^{12} years, in the 3:2 resonance. This would explain the observed difference in the distribution of asteroids in the 2:1 resonance (where only small asteroids are found) and 3:2 (where thousands are known, including some large ones). There was enough time for the escape of the large asteroids that primitively could be located in the 2:1 resonance, but there was no time for the same to happen in the 3:2 resonance.

The study of the same problem with the techniques of frequency analysis is an important alternative to corroborate the results found. The following figures (Fig. 3.12) are based on David Nesvorny and Sylvio Ferraz-Mello's results (Nesvorný and Ferraz-Mello 1997). The first figure (a) is a map of the 2:1 resonance. It is a composite of two complementary analyses. In both cases, the solutions were computed for over 1000 points, in three successive Δt intervals, using as a model the spatial motion of an asteroid in a system formed by the asteroid, the Sun, Jupiter, and Saturn. The frequency of the motion of the asteroidal perihelion f_ϖ was calculated first for the solution data between 0 and $2\Delta t$ and second for the solution data between Δt and $3\Delta t$. The two intervals in which the frequency was computed have an overlap, and their mean points are distant from each other by Δt. The plotted quantity is $\log_{10}|(f_2 - f_1)/f_1|$. In the upper part, the grid was formed by 1581 points (31×51) with $\Delta e = 0.01$ and $\Delta a = 0.004683$ AU; $\Delta t = 200{,}000$ years. In the lower part, a tighter grid was used. There are 2121 points (21×101) with $\Delta e = 0.01$ and $\Delta a = 0.001821$ AU; $\Delta t = 133,333$ years. Common parameters for both grids are the initial resonance angle $\theta = 0$ and the positions of the perihelion and the asteroidal node, aligned with Jupiter at the initial instant. The initial inclination of the asteroid is 0 (but that of Jupiter is the real one, as the model is three-dimensional). Four-body computations are extremely long, and it is not practical to build a dense network of points allowing the kind of representation used in the previous figure. We thus used a map of level curves of the values of $\log_{10}|(f_2 - f_1)/f_1|$, with the use of colors to better identify them.

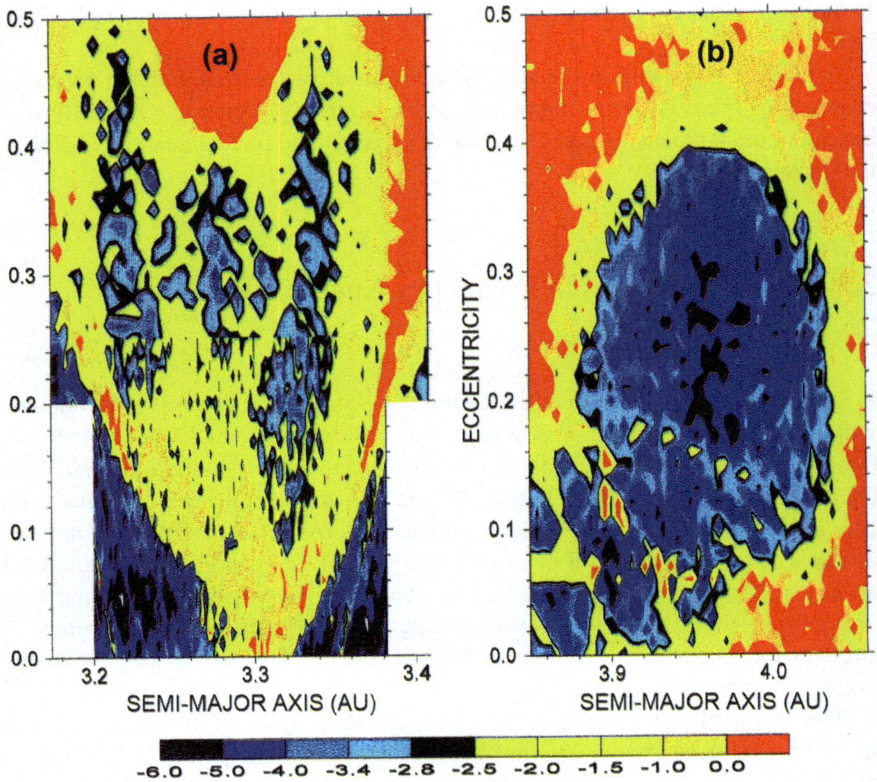

Fig. 3.12 Frequency variation maps of the 2:1 (**a**) and 3:2 (**b**) asteroidal resonances. The plotted parameter is the decimal logarithm of the relative variation of the perihelion angular velocity at 130 and 200 thousand years (see text). (Reprinted from Ferraz-Mello et al. (1998) with permission of the Astronomical Society of the Pacific.)

Similar diagrams were made for the 3:2 resonance, with 2601 points (51×51) with $\Delta e = 0.01$ and $\Delta a = 0.004162$ AU; $\Delta t = 200{,}000$ years and for the 4:3 resonance with 1804 points (41×44) with $\Delta e = 0.01$ and $\Delta a = 0.004683$ AU; $\Delta t = 200{,}000$ years.

These figures confirm the result obtained with the Lyapunov times. The chaoticity in the 2:1 resonance is much more intense than in the 3:2 resonance. All red and yellow areas in the figure correspond to areas of great variation of the frequency f_{ϖ}. The areas in various shades of blue, on the other hand, correspond to those in which the frequency variation is minimal, and a significant variation is not attainable in 4.5 billion years. A simple calculation using the rules of a random walk gives as result that, in the areas in lighter shades of blue, the frequency variations at 4.5 billion years are important, but without a more sophisticated diffusion model, we cannot say that diffusion will take the asteroid out of resonance.

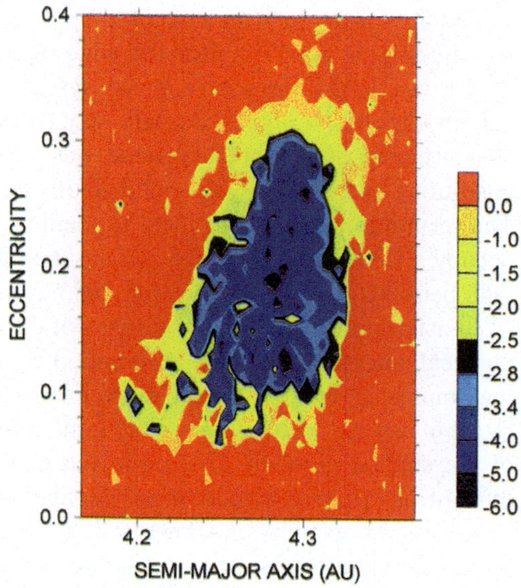

Fig. 3.13 Frequency variation map of the 4:3 asteroidal resonances. The plotted parameter is the decimal logarithm of the relative variation of the perihelion angular velocity at 200 thousand years.(Reprinted from Ferraz-Mello et al. (1998) with permission of the Astronomical Society of the Pacific.)

The same codes were used for the figure of the 3:2 resonance (Fig. 3.12b). While in resonance 2:1, the blue areas appear as a collection of tenuously connected patches, in resonance 3:2, the whole central area where the Hildas are located is blue, and even with darker shades indicating very stable motions. In resonance 4:3 (Fig. 3.13), the appearance is very similar to that of resonance 3:2 except that the blue area is much smaller than in resonance 3:2, and the frequency variation within it is as low as in resonance 3:2 (see Nesvorný and Ferraz-Mello 1997).

The only known large asteroid in the 4:3 resonance is (279) Thule, an asteroid more than 100 km in diameter. The non-existence of a larger number of large asteroids, despite the low diffusion rate, may be due to disturbances from the other giant planets: Uranus and Neptune, not considered in the used model.

3.9 Dynamical Maps on Dense Grids. Trojan Asteroids

A great advantage of the strategy used in the applications of frequency analysis and the dynamical maps of the previous sections is the coverage of dense grids of possibilities with hundreds, or even thousands, of points. Several parameters have been successfully used in maps over dense grids of initial conditions. Mathematicians have proposed maps based on the solution of the system formed by the equations of motion and their variational equations by replacing the limit of the calculation of the Lyapunov characteristic exponents by finite sums of chosen functions, the so-called FLI (Fast Lyapunov Indicators) discussed in Sect. 2.16 (see Froeschlé et al. 2000; Darriba et al. 2012; Maffione et al. 2011; Contopoulos 2002, sec. 2.10).

Another example is the map used by Rudolf Dvorak et al. (1998) in the study of the sticking of solutions near the edge of regular regions to identify the position of the last KAM torus of a given system. In this case, orbits that start in the neighborhood of a regular domain may remain attached to it for a very long time. The structure of this neighborhood was studied by taking a dense grid of initial conditions and counting the number of consequents before the solution escapes. The resulting map allowed them to identify the location of cantori (solutions that occupy a Cantor set on a torus and not the full torus) and to associate each of them with periodic orbits defined by rational numbers.

In astronomical applications, maps of dynamical characteristics with immediate physical interpretation had a certain preference. It is the case of the maps of maximum eccentricity and the maps of variation of the semi-major axis. However, by their very nature, these maps are influenced by the subjacent dynamical structure of the studied problem. If the adopted evolution times are not long enough, the dynamical structure will dominate the map and completely overshadow the chaotic evolution. An example is shown in Fig. 3.14. In this case, the quantity mapped in the third dimension is the maximum eccentricity reached by Jupiter in 10 million years in the vicinity of the 5:2 resonance. The definition of eccentricity mixes two independent momenta, and this map is not able to show the Arnold lattice with the finesse of other maps previously shown. But it immediately gives the interpretation of the phenomenology surrounding Jupiter's orbit in the resonance neighborhood, and this is a basic element for the astronomical study of this phenomenon.

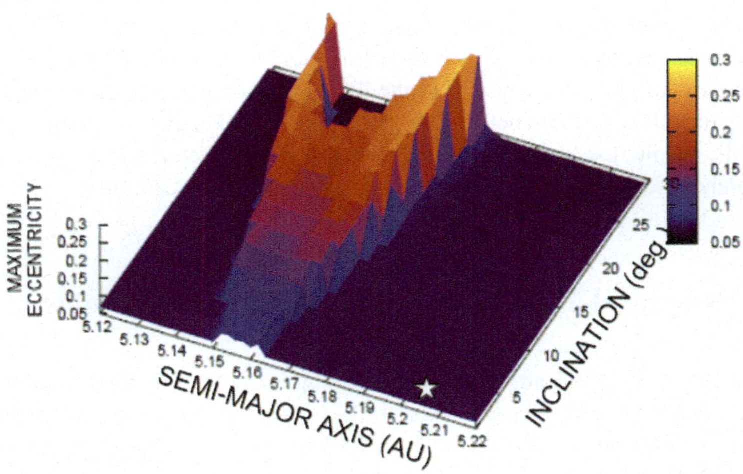

Fig. 3.14 3-D maps of the Jupiter-Saturn 5:2 resonance in the spatial case. The axes of the grid are the semi-major axis of Jupiter and the inclination of its orbital plane. The quantity mapped in the third dimension is the maximum value reached by the eccentricity in 10 million years. The maximum eccentricity corresponding to the current position of Jupiter is represented by a white star. © Rudolf Dvorak (Authorized reproduction)

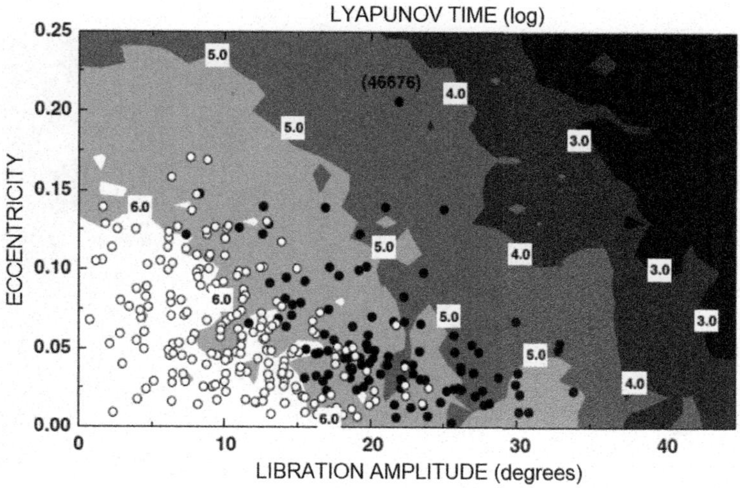

Fig. 3.15 Map of log T_{Lyap} for Trojan asteroids with an orbital inclination between 5 and 15°. The grid of initial conditions is formed by 31 × 26 points. The then-known real Trojan asteroids are represented by open dots and closed dots according to their Lyapunov times greater or less than 400,000 years, respectively. Times in years. (Reprinted from Tsiganis et al. (2005b). Copyright 2005 Springer.)

Theoretically, similar maps can be made using the LCE or the Lyapunov times. But from a practical point of view, it is impractical to do so due to the large computation time required for each initial condition. For instance, in the case of resonant asteroids, the dynamical maps of frequency variations were obtained with integrations over 200,000 years, while the Lyapunov times required integrations over 5 or 10 million years. One solution to this difficulty is to limit the calculation to a smaller sample of points. But even so, a number of points that is minimally reasonable can require immense computation times.

A remarkable example is shown in Fig. 3.15. There, the level lines are calculated from integrations starting at all 806 grid points for 4 million years and consider the gravitational attractions of the Sun and the four large planets. The not smooth appearance of the level lines is due to the typical indefinition in the determination of the Lyapunov times (see Figs. 2.28 and 2.36), which is as large as the value of T_{Lyap} is large and also to the existence of a large number of resonances going across the stable region of the studied grid. The integrations leading to $T_{\mathrm{Lyap}} > 400{,}000$ years (stable) were extended to 1 billion years to allow for greater accuracy in determining the higher Lyapunov times, and also the higher escape times shown in Fig. 3.16.

In the two graphs shown, the coordinates are the forced eccentricity (which coincides with the mean eccentricity if the eccentricity is not too small), and the amplitude of the oscillation (libration) of the asteroids' longitudes around the stationary Lagrange positions located at 60° from Jupiter (moving ahead of or behind Jupiter). The orbital inclinations are between 5° and 15°. Figures for other values of the inclination, up to 45°, can be found in Tsiganis et al. (2005b). The

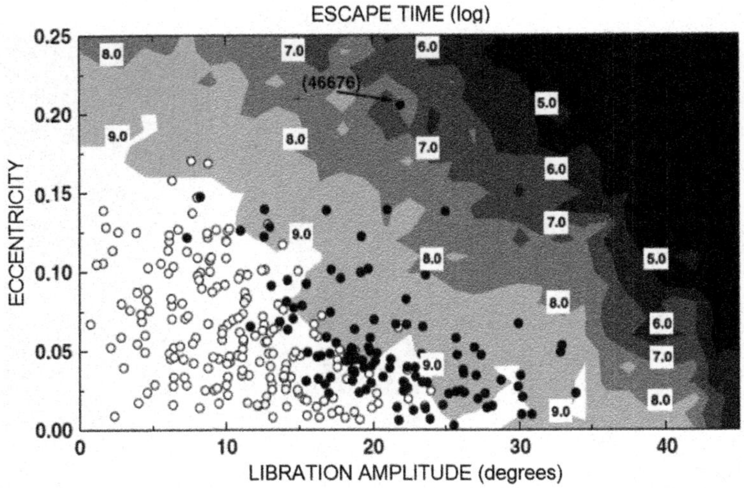

Fig. 3.16 Map of $\log T_{esc}$ for Trojan asteroids with an orbital inclination between 5° and 15°. Same description as in Fig. 3.15. (Reprinted from Tsiganis et al. (2005b). Copyright 2005 Springer.)

Lyapunov time and the escape time are approximately related by a power law (see Fig. 2.29). Note that 14% of the chaotic asteroids studied in Tsiganis et al. (2005b) have escape times $T_{esc} < 1$ Gyr. On the other hand, 20% of the chaotic asteroids have escape times longer than the age of the Solar System, i.e., 4.5 Gyr. These proportions are independent of the size of the asteroids, making it clear that its behavior is directly due to the resonances within this domain without interference from non-gravitational, i.e., thermal or collisional, phenomena.

An additional difficulty in constructing maps of Lyapunov times is the disparity of integration times needed to obtain them, which does not allow a priori bounds to be set for the integration times. The Lyapunov characteristic exponents are limits for $t \to \infty$ of the Oseledec function $\lambda(\mathbf{x}, \mathbf{e}, t)$ (see Sect. 2.11) and should not be confused with this function, whose behavior, moreover, can be quite erratic, as shown in Figs. 2.28 and 2.36.

The practice may impose considerations far from the concepts established in the mathematical theory of LCEs, but they should be present in the analysis of the results. For instance, in the Trojan study reported here, results in which the Lyapunov time was greater than 10% of the total integration time were not accepted. It is not valid to construct maps for values of the Oseledec λ-function for limited times t and use them to diagnose chaos. However, the use of the λ-function with a small integration time has been shown to be useful to map manifolds of fast transport in notoriously chaotic systems (see Fig. 2.31).

3.10 The Planets of the Pulsar PSR B1257+12

The discovery, in 1992, by Alexander Wolsczan and Dale Frail, of 3 planets in orbital motion around a pulsar surprised the astronomical community. First, by the fact that they were found orbiting the collapsed remnant of a star, and not a normal star, then by the fact that 3 planets were found, one of them comparable in size to the Moon. The first two planets discovered had masses respectively equal to 3.8 and 4.1 times the mass of the Earth, while the smallest of the 3 planets is still today one of the smallest, if not the smallest, ever discovered.

An important discovery was that the two planets with the greatest mass, PSR B1257+12 c and PSR B1257+12 d, complete one revolution around the pulsar in 66.5419 and 98.2114 days, respectively, that is, the orbital periods of the two planets are almost commensurable and very close to a 3:2 planetary resonance. This condition was fundamental for the confirmation of the discovery. Indeed, the event that precedes the collapse of the core of a star is the formation of a supernova, a violent explosion that can expel into space up to 90% of the surrounding matter. Pre-existing planets would not survive such a cataclysm. Other interpretations for the variations in the arrival times of pulsar pulses have been investigated. But the near resonance of the two orbits provided information allowing the discovery made to be confirmed.

When two planets move close to a resonance, the gravitational attraction between them gives rise to a perturbation in their motions of the same frequency as the resonant combination of their average motions (i.e., average angular velocities) and amplitude inversely proportional to the square of this frequency. This was discovered by Laplace in 1786 and served to remove the discrepancies which then existed between the results of the theories and the observed motion of the planets Jupiter and Saturn. In that case, the resonant angle originates from the 5:2 quasi-resonance between Jupiter and Saturn. In the case of the planets of PSR B1257+12, the resonance is different, but the phenomenon is the same. The two planets have periods very close to an exact commensurability, and the combination $(2\lambda_c - 3\lambda_d)$ of their longitudes has a frequency close to zero: $-3.0794 \times 10^{-3} \mathrm{d}^{-1}$ (or $-0.1764°/\mathrm{d}$), which corresponds to having an oscillation in the planets longitudes with a period of 5.59 years.

Despite the low frequency of the angle $(2\lambda_c - 3\lambda_d)$, the resulting oscillation in the longitudes is not huge because the masses of the two planets are very small. This is, however, compensated by the fact that the observations of the arrival times of pulses from a pulsar have an accuracy of less than one thousandth of a second. As a consequence, observations of the displacements of a pulsar with respect to the center of gravity of the planetary system it hosts are made with much higher precision than those obtained with normal star observation methods. The analysis of 3.5 years of observations of incoming pulses at the large radio telescope in Arecibo (Puerto Rico) allowed the detection of the oscillations due to the mutual gravitational attraction of the two planets and the determination of their actual masses and also, from these,

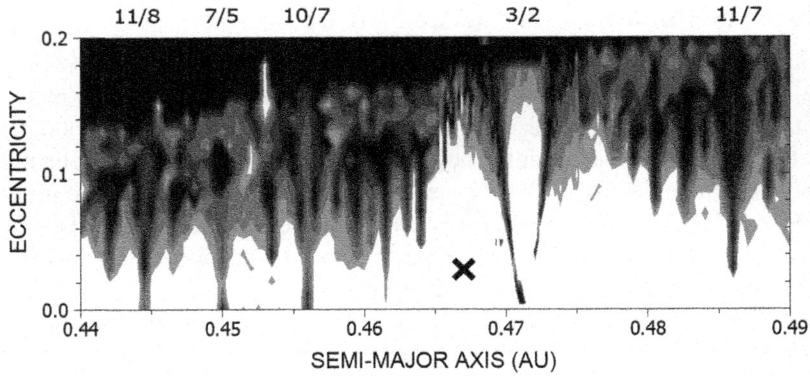

Fig. 3.17 Dynamical map of the surroundings of planet PSR B1257+12 d (represented by ×). The chaoticity measured by the spectral number N is represented with the same scale as in the figures of Sect. 3.3. The light regions are those of regular motion, and the dark regions are those of chaotic motion. The main resonances responsible for chaotic motion are indicated in the upper part of the figure. (Reproduced from Ferraz-Mello and Michtchenko (2002) with permission from Rev. Mexicana Astronomia y Astrofísica.)

the inclinations of the orbital planes of the two planets with respect to the line of sight of the pulsar (about 40°).

The good quality with which the orbital elements and the masses of the two largest planets of this system are known allows a reasonable determination of the domains in which these motions are chaotic, in the same way as was done for the large planets of the Solar System (see Sect. 3.3). Figure 3.17 shows the dynamical map of the surroundings of planet PSR B1257+12 d on a grid with 101 values of the semi-major axis and 21 values of the eccentricity. In this map, the chaoticity is measured by the spectral number N represented with the same scale as Fig. 3.2. The map shows extensive zones of regular motion at low eccentricities (blank areas in the lower part of the map), within one of which the planets of the pulsar are located. The various regular regions are separated by narrow bands of chaotic motion located at those values of the semi-major axis that correspond to resonances between the planets PSR B1257+12 c and PSR B1257+12 d. The main resonances responsible for chaotic motions are indicated in the upper part of the figure. Among the several visible resonances, the 3:2 resonance appears with prominence. It is formed by two chaotic branches in V corresponding to the separations between resonant and non-resonant motions, with a white area inside. It is not possible to say whether the motions in this white area are really regular or whether, as in the case of the asteroids in 3:2 resonance with Jupiter, they are chaotic motions with Lyapunov times so large that they escape detection in the limited time (100,000 years) of the simulations used to construct the map.

3.11 Exoplanets

Several projects to search for planets around nearby stars were carried out in the second half of the last century. Some discoveries were even announced, but it was not until 1995 that a discovery was confirmed. In that year, Michel Mayor and Didier Queloz, from the Astronomical Observatory of the University of Geneva (Switzerland), discovered a planet orbiting around the star 51 Pegasi. This was the first confirmed discovery of a planet in orbit around a Sun-like star, and since then, the number of exoplanets discovered and confirmed has grown exponentially (Fig. 3.18). The discovery of planet 51 Peg b opened a new chapter in the history of science and led the Royal Swedish Academy of Sciences to award the two astronomers the 2019 Nobel Prize in Physics. There are currently more than 5000 known exoplanets, and this number keeps increasing. Planets of various types have been discovered: sub-Earths, Earths, super-Earths, Neptunes, Jupiters, and super-Jupiters. Brown dwarf stars have also been found orbiting like a planet around normal stars. Although brown dwarfs are different from planets because of the occurrence of thermonuclear fusion reactions of Deuterium (^2H) at their centers, they have characteristics similar to those of super-Jupiters, and the separation of the two groups is poorly defined. Many authors differentiate them by masses, setting the boundary between super-Jupiters and brown dwarfs at 13 times the mass of Jupiter. The Encyclopedia of Extrasolar Planets, whose catalog was used to make Fig. 3.18, does not make any distinction between these planetary companions of great mass, limiting itself only to imposing the upper limit of 80 times the mass of Jupiter. From an observational point of view, in the transition zone between them, it is difficult to distinguish planets and brown dwarf stars. The differences are subtle, and only

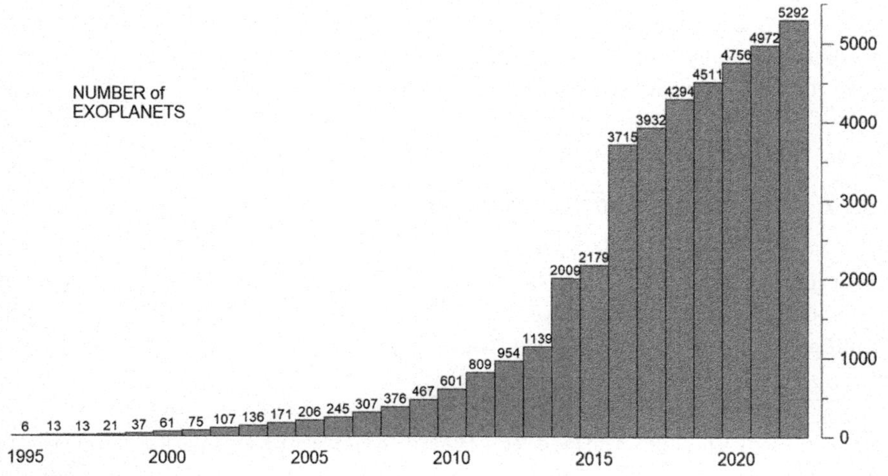

Fig. 3.18 Evolution of the number of known planetary companions from 1995 to 2022. Source: The Extrasolar Planets Encyclopaedia. http://exoplanet.eu/catalog/

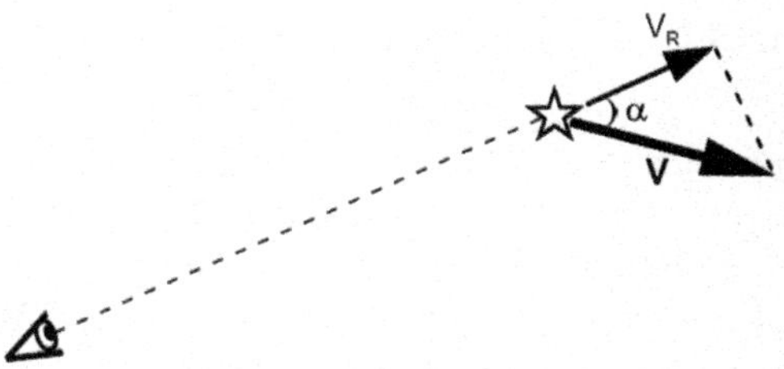

Fig. 3.19 Radial velocity of a star ($V_R = |\mathbf{V}| \cos \alpha$)

direct observations can identify them. From the dynamical point of view, they do not present differences, and the brown dwarfs acquire special importance because gravitational effects are proportional to the masses, and the masses of brown dwarfs are much greater than those of the planets.

An important fact is that 1/3 of the known planets belong to systems with at least two and up to eight planets. They are therefore, suitable for studies of resonances and chaos. However, these studies are hampered by inaccuracies in the determination of their orbital elements; in particular, errors in eccentricities can be large, and we see in the many examples of chaos and resonance shown in this book that differences in eccentricities can completely change the characteristics of a motion. But this is not the biggest obstacle. Larger systems, with orbital periods lasting months or years, are usually studied by measuring the radial velocity of the central star with respect to the system's barycenter. However, the Doppler effect reports only the value of the projection of the velocity of the star on the line of sight (Fig. 3.19), and this is not sufficient to determine the actual mass of a planetary companion, but only the so-called "minimum mass" given by the product $m \sin I$ where m is the real mass of the planet, and I is the inclination of the planet's orbital plane relative to a plane perpendicular to the line of sight of the star.[5]

The actual mass of the planet may be much greater than the "minimum mass" determined from radial velocities, and this difference has to be considered when studying the stability of an extrasolar planetary system whose parameters have been determined from measurements of radial velocities.

3.11.1 Example: Upsilon Andromedae

Consider as an example the first discovered planetary system formed by 3 planets around the star υ Andromedae with periods 4.6, 240.9, and 1281.4 days, and minimum masses 0.62, 1.8, and 10.2 times the mass of Jupiter, respectively.

[5] The same problem occurs with the determination of the planetary mass from the variations in the arrival times of the pulses of a pulsar. In the case studied in Sect. 3.10, the actual masses of the two major planets of the pulsar PSR B1257+12 were determined, as described, from perturbations in the longitude of the two planets, and the comparison of the values thus obtained with the minimum masses allowed the determination of the inclinations of their orbital planes.

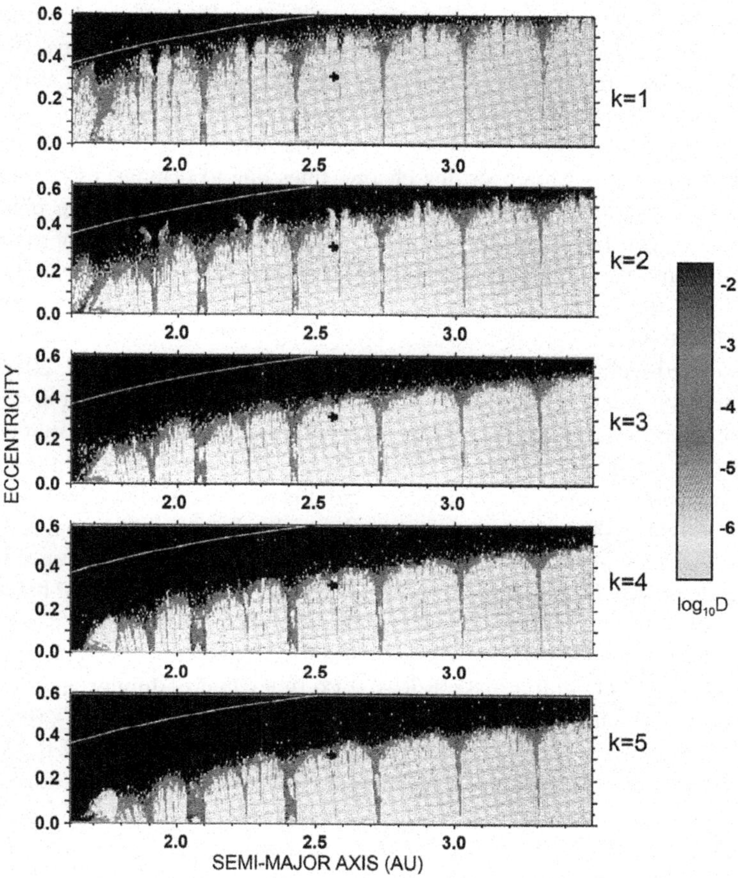

Fig. 3.20 Frequency variation maps of the vicinity of planet υ And d. The quantity plotted is the decimal logarithm of the relative variation in the average angular velocity of the planet in 50 thousand years. In the first map, at the top, the masses of the planets are the minimum masses determined by the observed radial velocities. In the other maps, the masses are multiplied by $k =$ 2, 3, 4, and 5, respectively. The position of the planet in the center of the maps is indicated by a cross. The white line indicates conditions in which the orbits of the two outer planets intersect. © Philippe Robutel (Authorized reproduction)

Figure 3.20 shows a set of frequency variation maps constructed by Philippe Robutel with the then-existing data. To take into account the lack of knowledge of the real masses, 5 maps were built using the minimum masses multiplied by 1, 2, 3, 4, and 5, respectively. These maps were constructed using the Laskar method of frequency analysis (see Sect. 3.8), keeping the orbital elements of the two inner planets fixed and varying the semi-major axis and the eccentricity of the third planet (υ And d). The current values of the elements of the third planet are indicated by a cross. The map constructed with the minimum masses ($k = 1$) shows that, in this case, the system would be currently in a very stable situation, with an eccentricity well below the domains of chaotic motion. It also shows that the chaotic lines due to resonant motions are thin and far away from the present system; the closest ones correspond to the 5:1 and 6:1 resonances of the two largest planets (Robutel and Laskar 2001).

In the maps built by multiplying the masses of the planets by a factor $k = 2, 3, 4, 5$, the boundary of the chaotic zone becomes lower and moves closer to the planet's position. Moreover, the chaotic lines due to resonances become more important. On these lines, large variations in the eccentricity of the planets can occur, which can take them to regions strongly affected by close approximations between them. The observation of this system with the Hubble Space Telescope showed that the orbits of the two outermost planets are in inclined planes with respect to the line of sight of the star.

Associated with the dynamical study of this system (Deitrick et al. 2014), these observations indicate for the orbital planes of υ And c and d, inclinations 11.3° and 25.6° relative to the plane tangent to the sky, respectively, and real masses much greater than the minimum masses. The values currently adopted are 9.1 and 23.6 times the mass of Jupiter. In this case, υ And d is not a planet but a brown dwarf.

The amount of difficulties surrounding the study of chaos in an exoplanet system is well illustrated by this example since, in addition to the great difference between the real masses and the minimum masses, we have a large inclination between the two orbital planes and the fact that the real mass of the planet υ And b is not known. We also have the unconfirmed existence of a fourth planet in the system, υ And e (Curiel et al. 2011), with a mass at least 1.06 times that of Jupiter and in an orbit well outside the other known planets, with a period of 3949.5 days (10.5 years). Without knowing the inclination of its orbit, we do not know its real mass, which can be quite large and capable of invalidating the results of analyses done without its exact consideration.

The set of difficulties reported in this example explains why, after a certain initial enthusiasm, astronomers have renounced very detailed studies of the chaos in extrasolar planetary systems, limiting themselves to a few examples and exploratory studies of some particular cases. To discover planets in extended orbits, even if they are Jupiters or super-Jupiters, is very difficult, and they are crucial in studies of chaos and stability.

3.12 MEGNO

A chaos indicator using the variational equations widely used, especially in the study of exoplanets, is the MEGNO (Mean Exponential Growth factor of Nearby Orbits) introduced by Pablo Cincotta and Carles Simò (2000). It is a technique that does not require as long integrations as the calculation of the Lyapunov characteristic exponents and allows us to verify whether a solution is regular or chaotic. This indicator is inspired by the fact that the Oseledec formula (see Sect. 2.11) used in the computation of the LCEs can be written as

$$\lambda(\mathbf{x}, t) = \frac{1}{t} \int \frac{\dot{\delta}(t)}{\delta(t)} dt \tag{3.1}$$

where $\delta(t) = ||\mathbf{e}(t)||$.

If we set the integration constant to $\ln ||e(0)||$ (which introduces the same difference in the calculated LCE as the formulas of Benettin et al.; see Sect. 2.11), we see that λ is the average, in the interval $[0, t]$, of the relative growth rate $\dot{\delta}/\delta$ (Cincotta and Giordano 2016). The MEGNO is a similar average but weighted by time (in fact, by $2t$):

$$Y(\mathbf{x}, t) = \frac{2}{t} \int_0^t \frac{\dot{\delta}(\tau)}{\delta(\tau)} \tau d\tau \tag{3.2}$$

which is used together with its average value in the considered interval:

$$\langle Y \rangle(\mathbf{x}, t) = \frac{1}{t} \int_0^t Y(\mathbf{x}, \tau) d\tau. \tag{3.3}$$

The MEGNO and its average value $\langle Y \rangle$ are dimensionless quantities. A simple calculation assuming that δ grows according to a power law ($\delta \sim \alpha t^n$) gives the result $\langle Y \rangle = 2n$. In the simple case of a Keplerian motion, as discussed in Sect. 1.1, δ will grow linearly with time, and hence $\langle Y \rangle = 2$. If δ varies exponentially ($\delta \sim e^{\alpha t}$), we obtain $\langle Y \rangle = \alpha t$, growing or decreasing linearly depending on whether one has $\alpha > 0$ or $\alpha < 0$. In more complex cases, the result can be more complex as well.

In general, the asymptotic behavior of $\langle Y \rangle$ is given by $\langle Y \rangle = at + b$ where $a = 0, b \simeq 2$, for regular motions, and $a = \frac{1}{2}\Lambda, b = 0$, where Λ is the maximum LCE, for chaotic motions. Because of the amplification of chaotic effects by the weighting introduced in the definition of $Y(x, t)$, MEGNO allows detecting signals of chaotic behavior much earlier than the LCE (Goździewski et al. 2001).

In practical applications, it is important to remember that numerical integrations are approximate operations that introduce errors in the results. These errors can accumulate faster than linear growth, which means that a residual propagation of numerical errors is always present in the solutions. In chaotic solutions, the effect is negligible, but in the case of regular solutions, they can affect the value of $\langle Y \rangle$ in a sensible way.[6]

[6] In integrations of planetary motions, rounding errors in the longitude, coordinates and velocities accumulate following a law in $t^{3/2}$ (Brouwer 1937; Beutler 2004).

3.12.1 Example: The Super Resonance of GJ 876

Let us consider as an example the study of the chaoticity of the three outermost planets of the star GJ 876 (GJ is the acronym of the Gliese catalog of stars close to the Sun). These three planets have masses 0.836, 2.66, and 0.05 times the mass of Jupiter. These masses are real and take into account the inclination of the orbital planes determined from the dynamical analysis in conjunction with measurements of radial velocities and astrometric observations made with the Hubble Space Telescope. They correspond to correcting the minimum masses by the factor $k = 1.25$ (Millholland et al. 2018).

The periods of these three planets are 30.1, 61.1, and 123.8 days, respectively. They are approximately commensurable two by two, and they are to each other like the numbers 1, 2, and 4. Their motion is close to a super-resonant periodic solution in which the periods are in the ratio 1:2:4. In this solution, the planets at a given moment are aligned along the axis of the innermost ellipse, on the side of their periastron (positions indicated with 1 in Fig. 3.21). After a complete revolution of planet 1 (30.1 days in the case of GJ 876), the other two planets will have moved to the positions indicated by 2 in the figure. After another complete revolution of 1, the other two planets will have moved to the positions indicated by 3 in the figure. They will be aligned again, but planet 3 will be on the other side. And so on. After four revolutions of planet 1, the three planets will return to alignment at positions 1. Note that when planets 1 and 2 (the most massive planets in the figure) are in alignment and on the same side, planet 1 is at the periastron of its orbit, making larger the distance between them in the alignment and decreasing their gravitational interaction. Note that if this had happened on the other side, with planet 1 in its apoastron, the two planets would be close to collision when aligned, and the solution could not be stable. In the same way, in their alignment, planets 2 and 3 are, respectively, at the periastron and apoastron of their orbits, thus maximizing their

Fig. 3.21 Periodic orbits in the 1:2:4 super-resonance

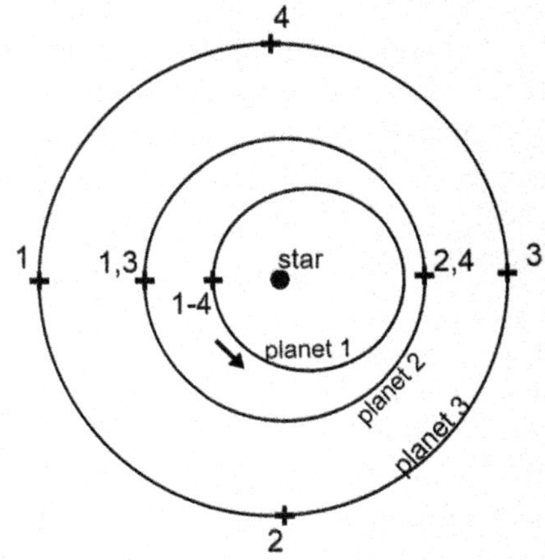

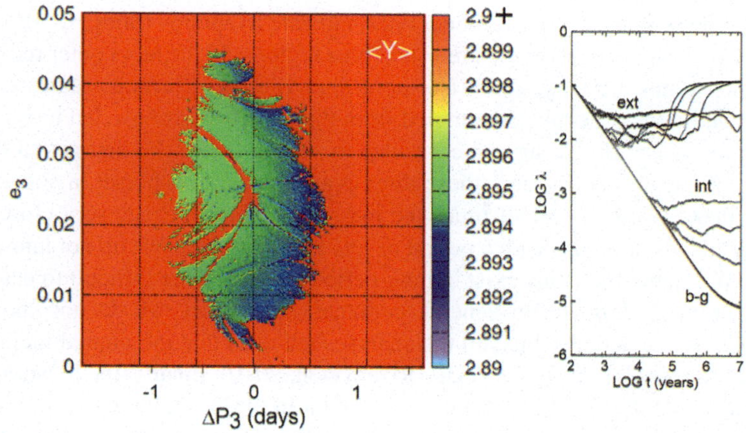

Fig. 3.22 *At Left*: Dynamical map of the surroundings of the 1:2:4 super-resonance of the system GJ 876. The quantity plotted is the MEGNO indicator $\langle Y \rangle$ calculated in a 100,000-year simulated evolution of the system. The adopted color scale enhances details of the immediate surroundings of the super-resonance. In the large red area, $\langle Y \rangle \sim 1000$. The abscissas indicate the period of the outer planet with respect to its value at the center of the figure. *On the right*: Calculation of the Lyapunov characteristic exponents. Log-log diagram of the functions $\lambda(t)$ (see Sect. 2.10) corresponding to solutions starting in the areas blue and green (b-g), the large outer red area (ext), and the red instability lines within the inner area (int). (Adapted from Martí et al. (2016). Copyright 2016 RAS.)

mutual distance. However, as the eccentricities of these two orbits are very small ($e \sim 0.03$), the orbits are almost circular, so this effect is not apparent in the figure.

Figure 3.22(left) shows a chart of the GJ 876 resonant system around the exact resonance as revealed by a numerical simulation of the 100,000-year evolution of the system. The MEGNO indicator $\langle Y \rangle$ was used. Its variation extends over several orders of magnitude, and the adopted color scale enhances details in the immediate surroundings of the super-resonance. In the red area, $\langle Y \rangle \sim 1000$. In this area, the motions are quite chaotic, and details are irrelevant. In the central area, where $\langle Y \rangle \sim 2.9$, the MEGNO value also corresponds to chaotic motions. The calculations of the Lyapunov characteristic exponents confirm this and show that it is a very low chaoticity. The log-log curves of $\lambda(t)$ (Fig. 3.22 right) obtained with initial conditions in the green and blue areas of the figure point to Lyapunov times exceeding 10^5 years (Martí et al. 2016). We do not know whether the power law of Lecar, Franklin and Murison (see Sect. 2.15) can be extended to a problem with the complexity of the planetary system of GJ 876. However, the result it gives is compatible with the age of the system, which can reach several billion years as indicated by the slow rotation of its central star (Rivera et al. 2005).

The central area of Fig. 3.22(left) shows a wealth of detail that can be associated with specific behaviors of the system solutions. The red lines correspond to zones of high chaoticity, probably associated with secondary resonances of the system. The calculation of the Lyapunov characteristic exponents with initial conditions taken on these lines, even the thinner ones, indicate Lyapunov times 10^3–10^4 years, much smaller than those of the green and blue regions. In the large outer red region, Lyapunov times are very small. No more than 10 years. Several studies using as initial conditions the orbits determined from the observations have led to very short Lyapunov times. This disagreement is due to several facts. The first is that the real system is not formed by only these three planets. A fourth planet, a super-Earth, with a mass seven and a half times the mass of the Earth, exists very close to the star and moves around it with a period of less than two days. We do not know about other planets, but they may exist in long-period orbits and are difficult to detect.

The used models are, in general, coplanar, but in fact, we do not know how coplanar are the actual orbits; all the analyses tell us is that the mutual inclinations between the larger planets are small (Rivera et al. 2010). Finally, the location of the blue-green stability region depends strongly on the values adopted for the various elements, mainly masses and eccentricities. We can therefore perform detailed studies, as in Fig. 3.22, but its real location may differ from the one we obtain. For this reason, we have changed the axis of the abscissae of the figure shown so that it shows the period P_3 relative to an arbitrary reference value that we place in the center of the figure. We may not know the exact value of P_3, but we know how much it can vary without changing the characteristics of the solutions.[7]

3.13 Compact Planetary Systems

The various space telescopes designed to search for exoplanets (CoRoT, Kepler, TESS) have resulted in a large number of new planets. The CoRoT telescope discovered the first compact system formed by super-Earths: CoRoT-7. The observations made by the Kepler telescope led to the discovery of more than 2600 planets, among them dozens of compact systems. The newest of these telescopes, TESS, in operation since 2018, is discovering many new systems.

In general, observations of planetary transits[8] allow us to determine the radius of the planets but not their masses. To determine the masses, it is necessary to observe the star with high-precision spectrographs installed in large telescopes on Earth. However, in the case of multiplanet systems, the high accuracy of the measurements made in space allows us, in some cases, to detect variations in the time elapsed between consecutive transits of a planet which are due to the motion of the star

[7] These final comments are not exclusive to the case studied. They are valid for all systems with many planets. It is often the case of solutions that well represent the observations but are unstable. MEGNO can be used to generate a penalty to be added to the quantity used for optimization in the process of fitting a solution to an observational data set and forcing it to stay in the most stable regions (Goździewski et al. 2006).

[8] Space telescopes measure the amount of light coming from the stars and detect the passage of planets in front of the star through the decrease in the amount of light measured. The decrease is as greater as greater is the radius of the planet transiting the star's disk.

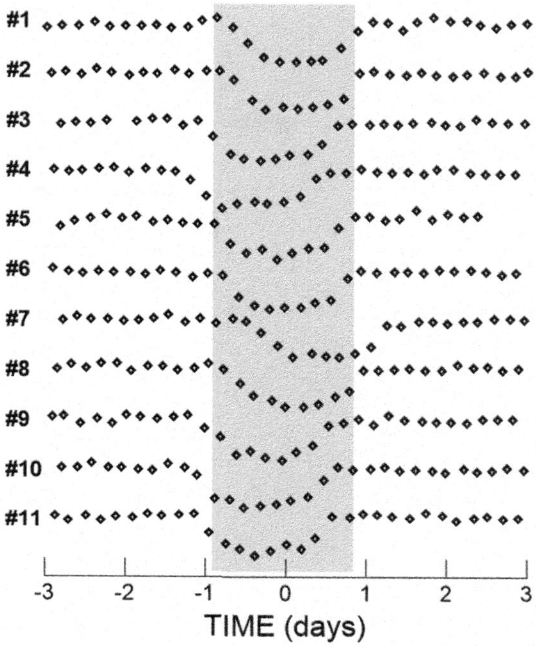

Fig. 3.23 Superposition of photometric observations of the star Kepler-46 on a sequence of transits of the planet Kepler-46 b through the star's disk. During the transits, 7–8% decays of the measured light quantity (flux) occur. The times between the transits are not always the same, and there are advances and delays of up to 25 minutes from the average (gray strip). Data from Nesvorný et al. (2012)

relative to the center of gravity of the system. As this motion depends on the masses of the planets in the system, its analysis based on the variation of the transits' time of occurrence (TTV) allows us to determine them.

Figure 3.23 shows the light curves of a sequence of transits of the planet Kepler-46 b, comparing them to a situation in which the transits would occur uniformly, with the center always in the middle of the gray strip. The first curves clearly show the transits occurring earlier and earlier, until curve #4, and then starting to be delayed, reaching the greatest delay at curve #7. These delays and advances can be up to 25 minutes and are due to an oscillation of the star relative to the center of gravity of the system, and are repeated every 190 days. It is worth noting that this variation is the only indication that the planet Kepler-46 c exists. Its orbit is inclined, and transits of this planet have not been observed. The best representation of the observed transit times is obtained with a model in which there are two giant planets—one Jupiter (Kepler-46 b) and one Saturn (Kepler-46 c)—with periods of 33.6 and 57 days, respectively, moving in nearly circular orbits around the planet.

The study of TTVs in the highly accurate observations of the Kepler space telescope not only allowed the discovery of some new planets not otherwise detectable, but, above all, allowed the determination of the masses in the many systems that showed multiple planets transiting the planet disk. Figure 3.24 shows planet systems with periods of less than 1 year discovered from observations by the Kepler space telescope, the vast majority of them in compact systems whose orbital periods do not exceed 60 days. The figure shows only planets whose masses have been determined.

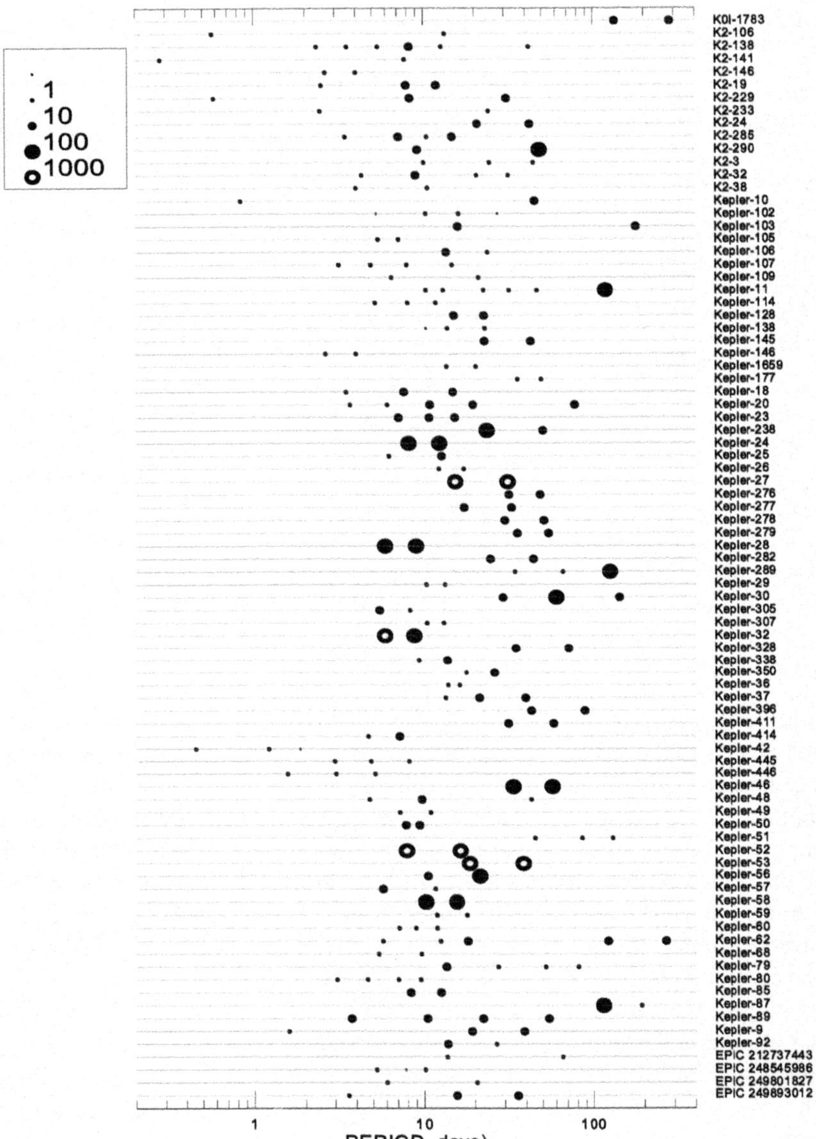

Fig. 3.24 Planetary systems with known masses and periods shorter than 1 year discovered with observations of the Kepler space telescope. The planets of each system are represented along a horizontal line according to their periods and by dots whose diameters indicate the order of magnitude of their masses distributed in 5 groups: $m < 1$; $1 < m < 10$; $10 < m < 100$; $100 < m < 1000$; $m > 1000$ times the mass of the Earth. Planets with no determined mass were not included. Source: The Extrasolar Planets Encyclopaedia. http://exoplanet.eu/catalog/

3.14 Resonant Chains

In systems with many planets, there can be cases of pairs of planets whose orbital periods are approximately commensurable, that is, in which the ratio of the two periods can be approximated by a simple rational number. In these cases, like the planets of GJ 876 studied in Sect. 3.12.1, the resonance phenomenon can manifest itself. As discussed in Sect. 1.19, due to the commensurability of the periods, mutual perturbations of the two planets can cause the existence of tubes of stationary periodic (or quasi-periodic) solutions (see Fig. 1.20), surrounded by separatrices that entangle themselves to form a dense chaotic layer separating these stationary domains from the rest of the phase space. A solution trapped in this tube is a resonant solution.

The above description is, in fact, a very simplified one. The paradigms that we use almost always reflect the reality in two degrees of freedom. In one problem with many more degrees of freedom, reality will instead be something like that shown in Fig. 3.22(left). The central zone (blue-green) is interspersed with instability lines (red). The edge is not well defined but rather appears to have a fractal structure. Another difference is that the central area resembles the expected stability region, but it is also a region of chaotic motion, yet weakly chaotic ($\langle Y \rangle < 3$). The only similarity with the simpler cases is the large red area of strongly chaotic solutions that surrounds the central area.

In general, it is only possible to say that the motions of two planets are in resonance after a complete dynamical study and when the orbital elements and masses are well determined. The first care is the correct definition of the critical angles. The forces of mutual attraction of the two planets do not depend on the reference system used. Therefore, the angles which appear in the expression of the forces should remain the same if the reference system in which the longitudes are measured is rotated. For this to be so, the sum of the coefficients of the angles which compose the critical angle must be equal to zero. In the case of the angle $\phi = k_1\lambda_1 - k_2\lambda_2$ (the λ_i are mean longitudes), the sum of the coefficients is $k_1 - k_2 \neq 0$. The real critical angle must therefore include, in addition to longitudes, the slow angles associated with the motions of the orbits. For example, in the case of the 2:1 resonance, the angle $\phi_j = \lambda_1 - 2\lambda_2 + \varpi_j$ (where the ϖ_j are the longitudes of the periastra).[9] A system of two resonant planets may have several critical angles.

For two planets to be in the resonance $k_2 : k_1$ ($k_2 > k_1$), at least one of the critical angles of this resonance must be oscillating around a fixed value (i.e., in libration). If all critical angles are circulating, the system is not in the resonance. Even so, the approximate commensurability of the periods is important. From classical theories, we know that the longitudes of the two planets have oscillations inversely proportional to the square of the composite angular velocity $k_1\dot{\lambda}_1 - k_2\dot{\lambda}_2$ (Tisserand 1889). If the two planets are very close to exact commensurability, this

---

[9] In the case of the Laplacian resonance of GJ 876, the critical angle $\phi = \lambda_1 - 3\lambda_2 + 2\lambda_3$ is invariant to rotations of the axes (since $1 - 3 + 2 = 0$).

Table 3.1 Resonant chains with 3 or more planets

Star/System	Planets known[a]	Chain	Notes
GJ 876 (see Sect. 3.12.1)	4 (1+3)	[2:1],[2:1]	
HD 158259	5	[3:2],[3:2],[3:2],[3:2]	Masses not known
HR 8799 (see Sect. 3.14.2)	4 (1+3)	[2:1],[2:1]	Extended
Kepler-60	3	[5:4],[4:3]	
Kepler-80	5 (1+4)	[3:2],[3:2],[4:3]	
Kepler-90 (2 chains)	8 (3+5)	[5:4],[5:3]	Masses not known
		[3:2],[4:3],[5:3],[11:7]	Masses not known
Kepler-223	4	[4:3],[3:2],[4:3]	Masses not known
K2-138	6 (5+1)	[3:2],[3:2],[3:2],[3:2]	
TOI-178 (see Sect. 3.14.1)	6 (1+5)	[2:1],[3:2],[3:2],[4:3]	
Trappist-1	7 (6+1)	[8:5],[5:3],[3:2],[3:2],[4:3]	
V 1298 Tauri	4 (3+1)	[3:2],[2:1]	Masses not known

[a] 1 means the existence of an inner or outer planet not involved in resonance with its known neighbours

velocity is very low, and the oscillation in the average longitudes of the planets will be very large. For instance, this large oscillation is the one detected in the system formed by the planets of the pulsar PSR B1257+12 (see Sect. 3.10), where it occurs thanks to the proximity of their periods to the 3/2 ratio.

The identification of a resonance becomes more difficult when it involves more than two planets, as in the resonant chains of Table 3.1. A classical example is the Laplacian resonance of the three innermost Galilean satellites of Jupiter: Io, Europa, and Ganymede. In this case, the mean angular velocities (the mean motions n_i) of the satellites are: 203.48, 101.37 and 50.32° per day, respectively, so that the combination $n_1 - 3n_2 + 2n_3$ is equal to zero (Ferraz-Mello 1979). In the usual notations used in texts about exoplanets, we would characterize this resonance as of type [2:1],[2:1] or 1:2:4. However, when we consider separately the pairs of satellites, the angles $\lambda_1 - 2\lambda_2$ e $\lambda_2 - 2\lambda_3$ circulate with velocity 0, 74°/d.

Let us first consider the pair Io-Europe. The main critical angles formed by the mean longitudes of these two satellites and the longitude of one of the periastra are

$$\begin{aligned} \phi_1 &= \lambda_1 - 2\lambda_2 + \varpi_1 \\ \phi_2 &= \lambda_1 - 2\lambda_2 + \varpi_2 \end{aligned} \tag{3.4}$$

Calculations with the classical theories (see Ferraz-Mello 1979, §6.4) showed that the first of these angles oscillates around 0 and the second around 180°. The first immediate consequence of these results is that the two periastra oscillate around permanently opposite positions on a line rotating in the retrograde direction with the same angular velocity as the angle $\lambda_1 - 2\lambda_2$, that is, 0.74°/d. In addition, every time

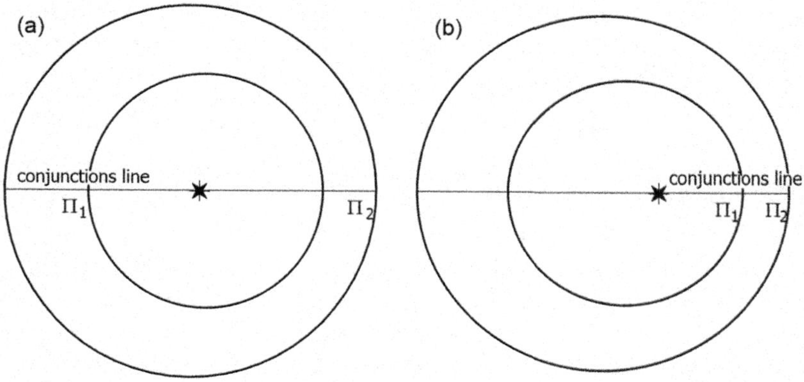

Fig. 3.25 2:1 resonance with apsidal corotation (ACR). (**a**) Case of two planets in anti-aligned orbits (the periastra of the two orbits, $\Pi_1$and Π_2, are on opposite sides of the line where the conjunctions of the two planets occur). (**b**) Idem with periastra aligned on the same side of the line of the conjunctions. (Reprinted from Ferraz-Mello et al. (2005). Copyright 2005 Springer.)

the planets line up on the same side of the planet (conjunctions), they do so on this same line.

These motions, which have been called resonances with apsidal corotation (ACR), are shown in Fig. 3.25. Panel (a) shows the case of the satellites Io and Europa: two very low eccentricity orbits (0.004 and 0.009) with periastra in permanent opposition. In panel (b), with two more sharply elliptic orbits, the periastra align on the same side. The detailed study of these resonances has shown the existence of much more complex cases in which the orbits intersect or in which the directions of the periastra form a fixed angle different from 0 or 180° (Beaugé et al. 2003; Ferraz-Mello et al. 2003, 2005) (see Sect. 3.15).

Since, due to the Laplacian resonance involving the first three satellites, the angle $\lambda_1 - 3\lambda_2 + 2\lambda_3$ oscillates around 180°, the resonance of Io and Europa implies the resonance of Europa and Ganymede. However, the dynamics of this second pair differs from that discussed above. The main critical angles formed by the average longitudes of these two satellites and the longitude of one of the periastra are

$$\begin{aligned} \phi_2' &= \lambda_2 - 2\lambda_3 + \varpi_2 \\ \phi_3' &= \lambda_2 - 2\lambda_3 + \varpi_3. \end{aligned} \tag{3.5}$$

The angle ϕ_2' has the same behavior as the angle ϕ_2 and oscillates around 180°. The angle ϕ_3', however, circulates, and the motion of the periastron of Ganymede is not dragged by the line of conjunctions like the periastron of the innermost satellites.

In several of the systems listed in Table 3.1 similar things happen. For instance, to make the commensurabilities of the periods of the K2-138 planets almost exact, we have to refer the motion of the planets to a rotating system with the speed of −2.5°/d. The same is true in several other cases. Pairs of planets in a super-resonant chain may or not be involved in a single resonance. This does not invalidate the presence of a super-resonance involving 3 or more planets, but it shows that the designation of resonance chains may not always correspond literally to reality.

3.14.1 *Example: TOI-178*

This first example is one of the super-Earth systems discovered by the TESS space telescope launched in 2018. The first results indicated the presence of some planets in the measurements made, but difficult to discern. In order to increase the available data, several observing campaigns were carried out with large telescopes and the CHEOPS space telescope. These additional observations showed that there were 6 planets which, with the exception of the first one, formed an intricate chain of super-resonances. The periods of these 5 planets (which we call P_1, P_2, P_3, P_4, P_5) have multiple commensurabilities and form 3 overlapping super-resonances characterized by the critical angles:

$$\begin{aligned} \phi &= \lambda_1 - 4\lambda_2 + 3\lambda_3 \\ \phi' &= 3\lambda_2 - 5\lambda_3 + 2\lambda_4 \\ \phi'' &= \lambda_3 - 2\lambda_4 + \lambda_5 \end{aligned} \tag{3.6}$$

relating the mean longitudes λ_i of these planets. The simulations conducted by the team that discovered the system showed that these critical angles oscillate around fixed values (librations). The periods determined show that the lines of the conjunctions of the four outer planets have a retrograde rotation with a speed of $\sim$1.38°/d. This is an indication that some periastra are being dragged by a line of conjunctions. More, however, cannot be said because the orbital eccentricities could not yet be determined. However, the dynamical study of the system shows that continuous regions of low chaoticity are concentrated at eccentricities very close to zero (see Fig. 3.26).

Figures 3.26 and 3.27 show two dynamical maps of the TOI-178 resonant chain obtained by frequency analysis (see Sect. 3.7). In both figures, the chaoticity is indicated by the variation of the mean angular velocity (Δn_5) of the most distant planet in two consecutive intervals of a simulation of the evolution of the system. The initial conditions correspond to the nominal orbits determined, except for the elements appearing in the axes of the maps, which were taken on a dense grid with the values indicated by the figures labels.

Figure 3.26 shows the chaoticity of the system in the vicinity of TOI-178 f. Almost the entire area shown corresponds to quite chaotic solutions. Only in a few points appear continuous areas of possibly regular solutions. The main one is exactly at the value of P_5 corresponding to the 2:3:4 super-resonance of the three outer planets of TOI-178. Other smaller ones lie in the center of structures characteristic of pairs of planets in resonance.

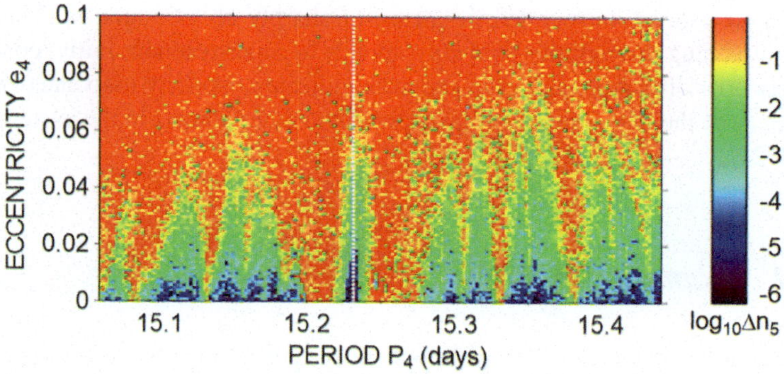

Fig. 3.26 Dynamical map of the surroundings of planet TOI-178 f obtained by frequency analysis. The chaoticity is measured by the variation of the mean angular velocity (n_5) of the most distant planet (according to the scale beside the figure). The value of the period P_4 determined from the observations is indicated by a white dotted line. (Reproduced from Leleu et al. (2021) with permission. © ESO.)

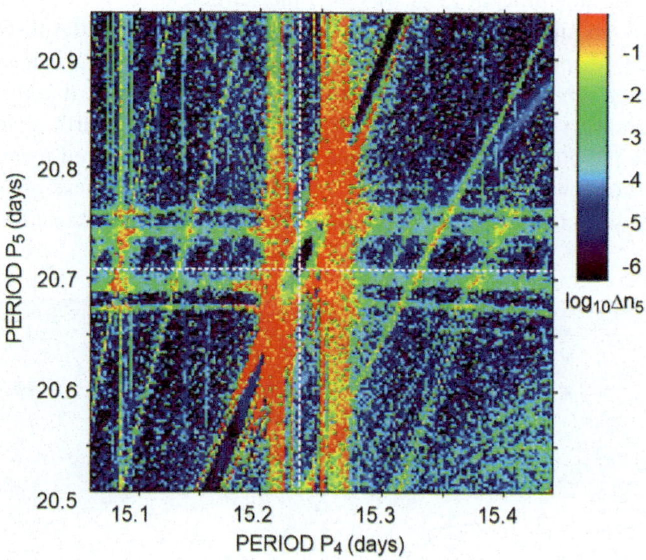

Fig. 3.27 Dynamical map of the surroundings of the 2:3:4 super-resonance of the outermost planets of TOI-178 obtained by frequency analysis. The quantity plotted is the mean angular velocity variation (Δn_5) of the most distant planet. The coordinates are the periods of the two outermost planets. The observed values of the periods are indicated by dotted white lines. (Reproduced from Leleu et al. (2021) with permission. © ESO.)

Figure 3.27 is a dynamical map of the surroundings of the 2:3:4 super-resonance of the outermost planets of TOI-178 as a function of the periods P_4 and P_5. The initial conditions are the nominal ones with all initial eccentricities equal to zero. The figure shows the expected patterns in this circumstance: An inner oval of more stable motion surrounded by a ring of sharply chaotic motion. The visible diagonal passing through this oval corresponds to super-resonant initial conditions. The networks of bands parallel to the axes indicate chaoticity associated with the resonances of the two outermost planets with the other planets in the system (Leleu et al. 2021).

3.14.2 Example: HR 8799

The planetary system HR 8799, with four super-Jupiters in extended orbits reaching 70 AU and orbital periods of up to 450 years, was the first multiplanet system detected by direct imaging (Fig. 3.28). The first images revealed only 3 planets, but as it became possible to see closer to the star, a fourth planet was discovered. Observations made over several years showed these planets moving. In addition, reprocessed archive images from the Hubble Space Telescope made it possible to determine their positions in 1998. With the measurements of their positions over 20 years, it was possible to determine their orbits and show that the three outermost planets have periods 115, 225, and 450 years, close to a 1:2:4 super-resonance. Their peculiar orbital architecture is still not well understood, mainly because so far we have observed only a short arc of their orbits. Of the fourth planet, which is smaller than the others and not seen in the observations recovered from the Hubble Space Telescope, we have data on only 10° of its orbit. Since these planets are very massive, between 6 and 9 times the mass of Jupiter (Goździewski and Migaszewski

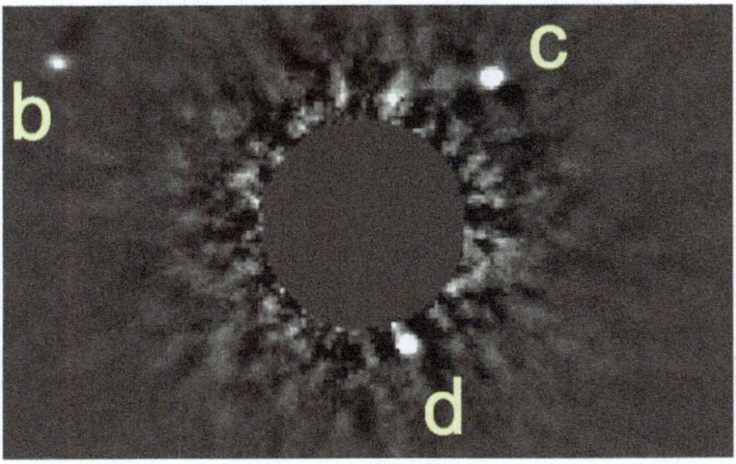

Fig. 3.28 Image showing the three outer planets of HR 8799 at the time of their discovery. The central region of the image is covered by an opaque circular disk. Credit: Gemini Observatory/NRC/AURA/Christian Marois et al. Sept. 2008

2020), the gravitational interactions are strong and the system may quickly become unstable if the orbits are not located in favorable conditions.

In the search for conditions of greater stability for this system, some authors have analyzed the observations of the planets conditioning the results to correspond to a 1:2:4:8 resonant chain involving the four planets (Goździewski and Migaszewski 2020; Wang et al. 2018).

The dynamical maps in the vicinity of these solutions are very similar to the one shown in Fig. 3.22(left) for the Laplacian resonance of GJ876, showing a central area of relative stability surrounded by a chaotic region. In this case, the semi-axis is larger than that obtained in unconditioned determinations and corresponds to a period of 482 years and a period ratio $P_4/P_3 = 2.76$, much larger than one would expect for the 2:1 resonance using the Kepler law. One of the characteristics of resonances with apsidal corotation is that the center of resonance moves away from the exact commensurability at low eccentricities (see Fig. 3.29) (Beaugé et al. 2003; Ferraz-Mello et al. 2005).

Astronomers often resort to a heuristic argument to postulate the stability of planetary systems: If systems were not in a stable situation, they would not have survived, in the formation they are in, for billions of years. But for HR 8799, this argument is not valid. This system is young, with the planets still immersed in the remains of the debris cloud from which they formed, and their age is estimated to be only 30 million years. It is important to continue analyzing new observations of this system to confirm whether it is indeed in a stable formation or whether the system is not stable and is evolving toward a catastrophic situation at a future time.

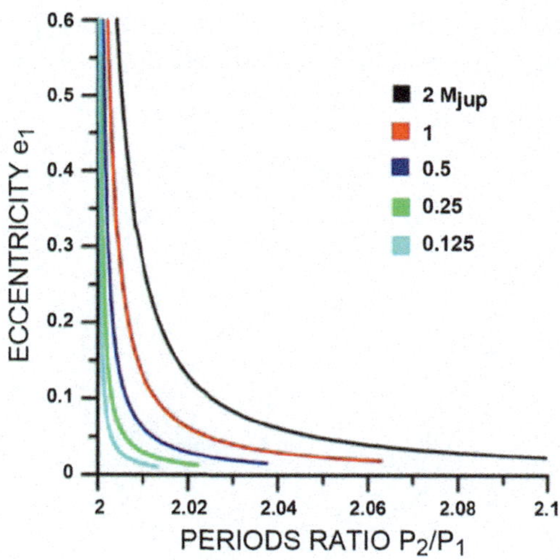

Fig. 3.29 Law of structure of the resonance 2:1. Values of the ratio of the periods as function of one of the eccentricities for pairs of planets of equal masses in apsidal corotation. (Reprinted from Michtchenko et al. (2008). Copyright 2008 RAS.)

3.15 Apsidal Corotation Resonance (ACR)

The examples presented in the previous sections show that the stability of resonant chains of exoplanets is associated with the proximity of their motions to stationary solutions. In the case of two planets, these solutions are known as apsidal corotation resonances (ACR) and were formerly studied by Cristian Beaugé et al. (2003, 2006), Sylvio Ferraz-Mello et al. (2003, 2006), and Tatiana Michtchenko et al. (2006). They can be symmetric or asymmetric. In the 2:1 resonance, the corotations occur as shown in Figs. 3.25 and 3.30. The eccentricities values are not any. There is a relationship between them that depends strongly on the mass ratio of the two planets so that, given one of them, the other will be defined, as well as the type of corotation (see Beaugé et al. 2006).

The ratio of the two periods also depends on the eccentricities and the masses. Figure 3.29 shows the values of P_2/P_1 (in the horizontal axis) in the case of the 2:1 resonance for two planets in orbit around a star of mass $1.15 M_{\odot}$, as a function of eccentricity e_1. The various curves correspond to two planets of equal masses in the range between 0.125 and 2 times the mass of Jupiter. These curves generalize, for the case of two planets, the Structure Law introduced in the 1980s in the study of asteroidal libration at high eccentricities (Ferraz-Mello 1988).

3.15.1 An Example of Capture and Evolution in the 2:1 Resonance

In the example shown in Fig. 3.31, two planets are initially in circular motions around the central star. An additional force increases the energy and the orbital

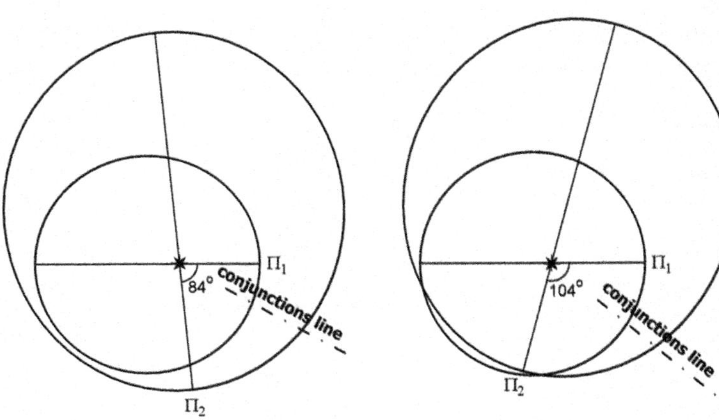

Fig. 3.30 2:1 resonance with asymmetric apsidal corotation (the periastra of the two orbits, Π_1 and Π_2 are not on the same line as in Fig. 3.25). *Left:* $|\delta\varpi| = 84°$, $e_1 = 0.286$ and $e_2 = 0.3$. *Right:* $|\delta\varpi| = 104°$, $e_1 = 0.17$ and $e_2 = 0.38$. (Reprinted from Ferraz-Mello et al. (2006). Copyright 2005 Springer.)

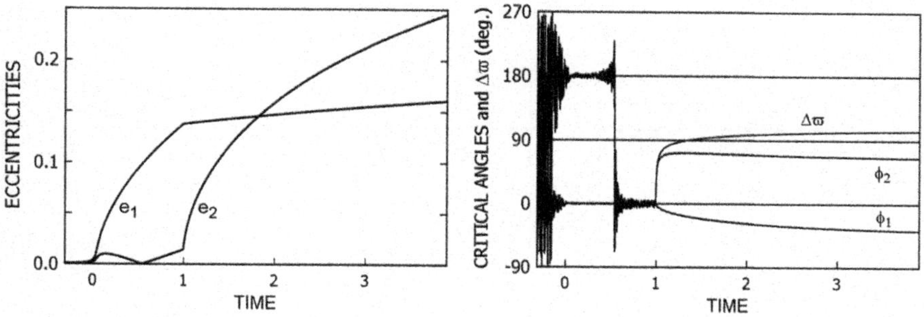

Fig. 3.31 Capture and evolution of a two-planet system in the 2:1 resonance from initial circular orbits. The system is captured in resonance and reaches, at $t = 0$, the stationary symmetric corotation solution, initially anti-aligned ($\Delta\varpi = 180°$), becoming aligned ($\Delta\varpi = 0$) at $t \sim 0.6$. Later, the corotation becomes asymmetric. The time scale was adjusted to give $t = 1$ in this last transition. (Adapted from Ferraz-Mello et al. (2003). Copyright 2003 Springer.)

angular momentum of the inner planet (this is easily simulated by assuming that the tide generated by the planet on the central star is deforming it and braking its rotation) so that the system approaches exact commensurability. Initially, the critical angles ϕ_1 and ϕ_2 (see Sect. 3.14) are circulating. When the system gets closer to the resonance, it is captured by it, and the critical angles begin to oscillate until they become fixed (at $t = 0$) at the stationary values $\phi_1 = 0$ and $\phi_2 = 180°$. The eccentricities then begin to vary. At $t \sim 0.6$, the orbit of the outer planet becomes circular, and its periastron changes sides when the orbit becomes elliptic again. The two periastra are now on the same side ($\Delta\varpi = 0$). A more important transition occurs at $t = 1$ when the corotation becomes asymmetric with the angle $\Delta\varpi$ increasing from 0 to values greater than 90°. The critical angles remain in libration but around values that vary over the evolution of the system.

Throughout the simulation of this example, the orbital eccentricity of the inner planet grows continuously. The periastra remain misaligned for a long time, but they come back into alignment when the internal eccentricity reaches $e_1 \sim 0.5$ (see Ferraz-Mello et al. 2003).

3.16 Dynamical Power Spectrum (Frequency Map)

Power spectra, as shown in Fig. 3.1, are graphs of the amplitude of the Fourier Transform as a function of frequency. To see how the spectra change when the initial conditions vary, it is necessary to construct many graphs and compare them. Representing them all together in a single figure is often an impossible task. To allow a global view of the spectral variations, Tatiana Michtchenko (Ferraz-Mello et al. 2003) introduced the so-called dynamical power spectrum (or frequency map) obtained by constructing the Fourier transforms corresponding to a sequence of

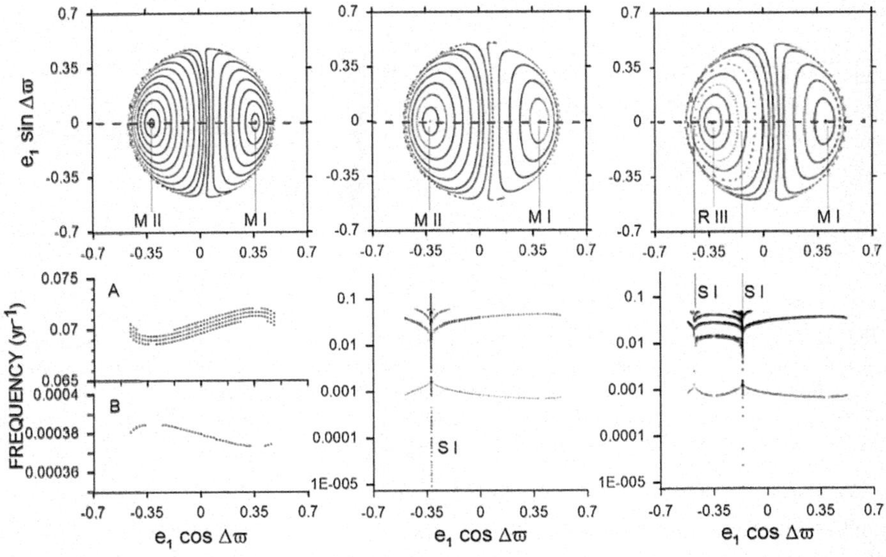

Fig. 3.32 Surfaces of Section and dynamical power spectra for 3 values of the energy in the vicinity of the 3:2 resonance of the system formed by planets b and c of the pulsar PSR B1257+12 when the averaged equations are considered. In the first panel, the dynamical power spectrum was divided into two parts (A and B) to clearly show the main frequencies despite the fact that they are very different. (Adapted from Callegari et al. (2006). Copyright 2006 Springer.)

initial conditions and marking on a graph of frequency vs. initial condition, all points where the amplitude of the Fourier transform is greater than a fixed threshold.

For instance, the lower part of Fig. 3.32 shows the dynamical power spectra of three sets of solutions whose initial conditions are taken on the center line of the Surfaces of Section shown in the upper part of the figure. These initial conditions are defined by $e_1 \sin \Delta\varpi = 0$, and the spectra are parametrized by the abscissas $x = e_1 \sin \Delta\varpi$.

If a solution is regular, its Fourier transform is like the one shown in Fig. 3.1a, with few peaks above the limit level. This is the case for all solutions in the first panel of Fig. 3.32. For each value of the abscissa, we have at most 5 points (4 in part A and 1 of low frequency in part B of the graph). The frequencies vary when x varies, but in a continuous way without irregularities.

If a solution is chaotic, its Fourier transform looks like the one shown in Fig. 3.1b, with a large number of peaks above the threshold level; This is what happens with the solutions indicated as type S1 in Fig. 3.32. The large number of frequencies with amplitudes above the limit level is represented in the dynamical spectrum by a large number of vertically distributed points at the abscissa corresponding to this solution. These structures indicate that the corresponding solutions on the Surface of Section are chaotic.

An intermediate situation occurs in the third panel of Fig. 3.32. Between the two vertical lines indicated by S1, the number of lines connecting the points corresponding to the values of x in this interval appear finite, but it is large with the lines thickened by the superposition of very close frequencies. This indicates that the motion in this region is much more complex than, for instance, the motions corresponding to the first panel. In the example shown, the planets have small masses (super-Earths), but if they were Jupiters with large masses, a band of chaotic motions would probably form between the two lines indicated by S1.

The lesson to be learned from this example is that the dynamical spectrum helps to identify features of the Surface of Section that are not visible in the graphical representation of the sections. The visual analysis of the three sections shown in the upper part of Fig. 3.32 does not allow us to see any major difference between them. They look like sections of regular motions for all initial conditions of the domain studied. But the dynamical spectra show that this is not so. In the first panel, the motions are indeed regular and are oscillations around two centers denoted by MI and MII where, respectively, $\Delta\varpi = 0$ and $\Delta\varpi = 180°$. In the following panels, the motion first becomes chaotic at the center MII, and then a ring-shaped separatrix is formed where the motions are chaotic. It separates a domain outside the ring, where the motions are regular, and a domain inside the ring, where the motions become much more complex (the number of frequency peaks increases greatly). In the example, the Surfaces of Section were selected to show a problem that can occur and how the dynamical power spectra serve to solve it. In order not to leave the information incomplete, it is necessary to add that, for other values of the energy, separatrices become evident on the Surfaces of Section of this problem (Callegari et al. 2006).

References

Beaugé, C., Ferraz-Mello, S., Michtchenko, T.A.: Extrasolar planets in mean-motion resonance: apses alignment and asymmetric stationary solutions. Astrophys. J. **593**, 1124–1133 (2003)

Beaugé, C., Michtchenko, T.A., Ferraz-Mello, S.: Planetary migration and extrasolar planets in the 2/1 mean-motion resonance. Mon. Not. R. Astron. Soc. **365**, 1160–1170 (2006)

Beutler, G.: Methods of Celestial Mechanics, vol. I, p. 339. Springer, Berlin, Heidelberg (2004)

Bohr, H.: Almost Periodic Functions. Chelsea Publ., New York (1951)

Brouwer, D.: On the accumulation of errors in numerical integration. Astron. J. **46**, 149–153 (1937)

Callegari Jr, N., Ferraz-Mello, S., Michtchenko, T.A.: Dynamics of two planets in the 3/2 mean-motion resonance: application to the planetary system of the pulsar PSR B1257+ 12. Celestial Mech. Dynam. Astron. **94**, 381–397 (2006)

Celletti, A., Chierchia, L.: KAM stability and Celestial Mechanics. Memoirs of the American Mathematical Society, No. 878 (2007)

Cincotta, P.M., Giordano, C.M.: Theory and applications of the mean exponential growth factor of nearby orbits (MEGNO) method. In: Skokos, C., et al. (eds.) Chaos Detection and Predictability, pp. 93–128. Springer, Berlin, Heidelberg (2016)

Cincotta, P.M., Simò, C.: Simple tools to study global dynamics in non-axisymmetric galactic potentials – I. Astron. Astrophys. Suppl. Ser. **147**, 205–228 (2000)

Contopoulos, G.: Order and Chaos in Dynamical Astronomy. Springer, Berlin (2002)

Correia, A.C.M., Laskar, J.: Tidal evolution of exoplanets. In: Seager, S. (ed.) Exoplanets. Univ. Arizona Press, Tucson (2010)

Curiel, S., Cantó, J., Georgiev, L., Chávez, C.E., Poveda, A.: A fourth planet orbiting υ Andromedae. Astron. Astrophys. **525**, A78 (2011)

Darriba, L.A., Maffione, N.P., Cincotta, P.M., Giordano, C.M.: Comparative study of variational chaos indicators and ODEs' numerical integrators. Int. J. Bifurc. Chaos **22**(10), 1230033 (2012)

Deitrick, R., Barnes, R., McArthur, B., Quinn, T.R., et al.: The three-dimensional architecture of the υ Andromedae planetary system. Astrophys. J. **798**, 46 (2014)

de Sousa Ribeiro, R., Morbidelli, A., Raymond, S.N., Izidoro, A. et al.: Dynamical evidence for an early giant planet instability. Icarus **339**, 113605 (2020)

Dumas, H.S., Laskar, J.: Global dynamics and long-time stability in Hamiltonian systems via numerical frequency analysis. Phys. Rev. Lett. **70**, 2975–2979 (1993)

Dvorak, R., Contopoulos, G., Efthymiopoulos, C., Voglis, N.: "Stickiness" in mappings and dynamical systems. Planet. Space Sci. **46**, 1567–1578 (1998)

Ferraz-Mello, S.: Dynamics of the Galilean Satellites. Inst. Astron. Geof. Univ. São Paulo, São Paulo (1979). http://www.astro.iag.usp.br/~dinamica/DGSX.pdf

Ferraz-Mello, S.: The high-eccentricity libration of the Hildas. Astron. J. **96**, 400–408 (1988)

Ferraz-Mello, S.: Canonical Perturbations Theories. Degenerate Systems and Resonance. Springer, New York (2007)

Ferraz-Mello, S.: Tidal synchronization of close-in satellites and exoplanets: II. Spin dynamics and extension to Mercury and exoplanets host stars. Celest. Mech. Dyn. Astron. **122**, 359–389 (2015). Errata: id. **130**: 78, pp. 20–21 (2018)

Ferraz-Mello, S., Michtchenko, T.A.: Extra-solar planetary systems. Rev. Mex. Astron. Astrofis. **14**, 7–10 (2002)

Ferraz-Mello, S., Nesvorný, D., Michtchenko, T.A.: Chaos, diffusion, escape and permanence of resonant asteroids in gaps and groups. In: Lazzaro, D., et al. (eds.) Solar System Formation and Evolution, pp. 65–82. Astron. Soc. Pacific, San Francisco (1998)

Ferraz-Mello, S., Beaugé, C., Michtchenko, T.A.: Evolution of migrating planet pairs in resonance. Celest. Mech. Dyn. Astron. **87**, 99–112 (2003)

Ferraz-Mello, S., Michtchenko, T.A., Beaugé, C., Callegari Jr., N.: Extrasolar planetary systems. In: Dvorak, R., et al. (eds.) Chaos and Stability in Planetary Systems, pp. 219–271. Springer, Berlin, Heidelberg (2005)

Ferraz-Mello, S., Michtchenko, T.A., Beaugé, C.: Regular motions in extra-solar planetary systems. In: Steves, B., Maciejewski, A. (eds.) Chaotic Worlds: From Order to Disorder in Gravitational N-Body Dynamical Systems, pp. 255–288. Springer, Dordrecht (2006)

Froeschlé, C., Guzzo, M., Lega, E.: Graphical evolution of the Arnold web: from order to chaos. Science **289**, 2108–2110 (2000)

Giorgilli, A., Locatelli, U., Sansottera, M.: Kolmogorov and Nekhoroshev theory for the problem of three bodies. Celest. Mech. Dynam. Astron. **104**, 159–173 (2009)

Gomes, R., Levison, H.F., Tsiganis, K., Morbidelli, A.: Origin of the cataclysmic Late Heavy Bombardment period of the terrestrial planets. Nature **435**, 466–469 (2005)

Goździewski, K., Bois, E., Maciejewski, A.J., Kiseleva-Eggleton, L.: Global dynamics of planetary systems with the MEGNO criterion. Astron. Astrophys. **378**, 569–586 (2001)

Goździewski, K., Konacki, M., Maciejewski, A.J.: Orbital configurations and dynamical stability of multiplanet systems around Sun-like stars HD 202206, 14 Herculis, HD 37124, and HD 108874. Astrophys. J. **645**, 688–703 (2006)

Goździewski, K., Migaszewski, C.: An exact, generalized Laplace resonance in the HR 8799 planetary system. Astrophys. J. Lett. **902**, L40 (2020)

Hénon, M.: Exploration numérique du problème restreint. IV. Masses égales, orbites non périodiques. Bull. Astron. 3e.série **1**, 49–66 (1966)

Laskar, J.: The chaotic motion of the Solar System: a numerical estimate of the size of the chaotic zones. Icarus **88**, 266–291 (1990)

Laskar, J.: Large scale chaos and marginal stability in the Solar System. Celest. Mech. Dynam. Astron. **64**, 115–162 (1996)

Laskar, J.: Marginal stability and chaos in the Solar System. In: Sympos. International Astronomical Union, vol. 172, pp. 75–88 (1996)

Laskar, J., Robutel, P.: The chaotic obliquity of the planets. Nature **361**, 608–612 (1993)

Laskar, J., Froeschlé, C., Celletti, A.: The measure of chaos by the numerical analysis of the fundamental frequencies. Application to the standard mapping. Phys. D Nonlinear Phenom. **56**, 253–269 (1992)

Leleu, A., Alibert, Y., Hara, N.C., Hooton, M.J., et al.: Six transiting planets and a chain of Laplace resonances in TOI-178. Astron. Astrophys. **649**, A26 (2021)

Maffione, N.P., Darriba, L.A., Cincotta, P.M., Giordano, C.M.: A comparison of different indicators of chaos based on the deviation vectors: application to symplectic mappings. Celest. Mech. Dynam. Astron. **111**, 285–309 (2011)

Martí, J.G., Cincotta, P.M., Beaugé, C.: Chaotic diffusion in the Gliese-876 planetary system. Mon. Not. R. Astron. Soc. **460**, 1094–1105 (2016)

Michtchenko, T.A., Ferraz-Mello, S.: Resonant structure of the outer Solar System in the neighborhood of the planets. Astron. J. **122**, 474–481 (2001)

Michtchenko, T.A., Beaugé, C., Ferraz-Mello, S.: Stationary orbits in resonant extrasolar planetary systems. Celest. Mech. Dynam. Astron. **94**, 411–432 (2006)

Michtchenko, T.A., Beaugé, C., Ferraz-Mello, S.: Dynamic portrait of the planetary 2/1 mean-motion resonance–I. Systems with a more massive outer planet. Mon. Not. R. Astron. Soc. **387**, 747–758 (2008)

Millholland, S., Laughlin, G., Teske, J., Butler, R.P., et al.: New Constraints on Gliese 876 - Exemplar of mean-motion resonance. Astron. J. **155**, 106 (2018)

Moisson, X., Bretagnon, P.: Analytical planetary solution VSOP2000. Celest. Mech. Dynam. Astron. **80**, 205–213 (2001)

Mojzsis, S.J., Brasser, R., Kelly, N.M., Abramov, O., Werner, S.C.: Onset of giant planet migration before 4480 million years ago. Astrophys. J. **881**, 44 (2019)

Neron de Surgy, O., Laskar, J.: On the long term evolution of the spin of the Earth. Astron. Astrophys. **318**, 975–989 (1997)

Nesvorný, D., Ferraz-Mello, S.: On the asteroidal population of the first-order jovian resonances. Icarus **130**, 247–258 (1997)

Nesvorný, D., Kipping, D.M., Buchhave, L.A., Bakos, G.Á., et al.: The detection and characterization of a nontransiting planet by transit timing variations. Science **336**, 1133–1136 (2012)

Noid, D.W., Koszykowski, M.L., Marcus, R.A.: A spectral analysis method of obtaining molecular spectra from classical trajectories. J. Chem. Phys. **67**, 404–408 (1977)

Rivera, E.J., Lissauer, J.J., Butler, R.P., Marcy, G.W., et al.: A 7.5 $M_\oplus$ planet orbiting the nearby star GJ 876. Astrophys. J. **634**, 625–640 (2005)

Rivera, E.J., Laughlin, G., Butler, R.P., Vogt, S.S., et al.: The Lick-Carnegie exoplanet survey: a Uranus-mass fourth planet for GJ 876 in an extrasolar Laplace configuration. Astrophys. J. **719**, 890–899 (2010)

Robutel, P., Laskar, J.: Frequency map and global dynamics in the Solar System I. Icarus **152**, 4–28 (2001)

Sussman, G.J., Wisdom, J.: Numerical evidence that the motion of Pluto is chaotic. Science **241**, 433–437 (1988)

Sussman, G.J., Wisdom, J.: Chaotic evolution of the solar system. Science **257**, 56–62 (1992)

Tisserand, F.: Traité de Mécanique Céleste, tome I. Gauthier-Villars, Paris (1889), sec. 74

Tsiganis, K., Gomes, R., Morbidelli, A., Levison, H.F.: Origin of the orbital architecture of the giant planets of the Solar System. Nature **435**, 459–461 (2005a)

Tsiganis, K., Varvoglis, H., Dvorak, R.: Chaotic diffusion and effective stability of Jupiter Trojans. Celest. Mech. Dynam. Astron. **92**, 71–87 (2005b)

Wang, J.J., Graham, J.R., Dawson, R., Fabrycky, D., et al.: Dynamical constraints on the HR 8799 planets with GPI. Astron. J. **156**, 192 (2018)

Correction to: Chaotic Dynamics in Planetary Systems

Correction to:
S. Ferraz-Mello, *Chaotic Dynamics in Planetary Systems*, Astronomy and Planetary Sciences, https://doi.org/10.1007/978-3-031-45816-3

This book was inadvertently published without updating the following corrections:

p. 25—line 7 from the top, "projectedy" has been updated as "projected".
p. 97—line 12 from the bottom, "varieties" has been updated as "manifolds".
p. 105—line 6 from the top, "orbits" has been updated as "curves".
p. 119—line 5 from the bottom, the sentence "frequency associated with he" has been updated as "frequency associated with the".
p. 147—line 3 from the top of text, "of variations (TTV)" has been updated as "of occurrence (TTV)".
p. 161—line 4 from the bottom, "(2005b)" from the reference Tsiganis, K., Varvoglis, H., Dvorak, R. has been deleted.
p. 163—last line, "Transit time of variations" has been updated as "Transit time variations".
p. 167—left side of column 1, 6th line from the bottom, "Separarix" has been updated as "Separatrix".

The updated version of this book can be found at
https://doi.org/10.1007/978-3-031-45816-3

S. Ferraz-Mello, *Chaotic Dynamics in Planetary Systems*,
Astronomy and Planetary Sciences, https://doi.org/10.1007/978-3-031-45816-3_4

Acronyms

ACR	Apsidal corotation resonance
AU	Astronomical unit
FFT	Fast Fourier Transform
FLI	Fast Lyapunov indicator
IMCCE	Institute of Celestial Mechanics and Calculation of Ephemerides
KAM	Kolmogorov–Arnold–Moser
LCE	Lyapunov characteristic exponents
MEGNO	Mean exponential growth factor of nearby orbits
NEA	Near-Earth asteroid
TTV	Transit time variations

S. Ferraz-Mello, *Chaotic Dynamics in Planetary Systems*,
Astronomy and Planetary Sciences, https://doi.org/10.1007/978-3-031-45816-3

Index

S. Ferraz-Mello, *Chaotic Dynamics in Planetary Systems*,
Astronomy and Planetary Sciences, https://doi.org/10.1007/978-3-031-45816-3

SPRINGER NATURE

GPSR Compliance

The European Union's (EU) General Product Safety Regulation (GPSR) is a set of rules that requires consumer products to be safe and our obligations to ensure this.

If you have any concerns about our products, you can contact us on ProductSafety@springernature.com

In case Publisher is established outside the EU, the EU authorized representative is:

Springer Nature Customer Service Center GmbH
Europaplatz 3
69115 Heidelberg, Germany

The manufacturer's authorised representative in the EU is Springer
Nature Customer Service Centre GmbH, Europaplatz 3, 69115 Heidelberg,
Germany. If you have any concerns regarding our products, please
contact ProductSafety@springernature.com

Printed and bound by CPI Group (UK) Ltd, Croydon, CR0 4YY
28/09/2026
02224085-0017